Applications of Membrane Technology for Food Processing Industries

Applications of Membrane Technology for Food Processing Industries

Edited by
M. Selvamuthukumaran

CRC Press is an imprint of the
Taylor & Francis Group, an **informa** business

First edition published 2021
by CRC Press
6000 Broken Sound Parkway NW, Suite 300, Boca Raton, FL 33487-2742

and by CRC Press
2 Park Square, Milton Park, Abingdon, Oxon, OX14 4RN

© 2021 Taylor & Francis Group, LLC

CRC Press is an imprint of Taylor & Francis Group, LLC

Reasonable efforts have been made to publish reliable data and information, but the author and publisher cannot assume responsibility for the validity of all materials or the consequences of their use. The authors and publishers have attempted to trace the copyright holders of all material reproduced in this publication and apologize to copyright holders if permission to publish in this form has not been obtained. If any copyright material has not been acknowledged please write and let us know so we may rectify in any future reprint.

Except as permitted under U.S. Copyright Law, no part of this book may be reprinted, reproduced, transmitted, or utilized in any form by any electronic, mechanical, or other means, now known or hereafter invented, including photocopying, microfilming, and recording, or in any information storage or retrieval system, without written permission from the publishers.

For permission to photocopy or use material electronically from this work, access www.copyright.com or contact the Copyright Clearance Center, Inc. (CCC), 222 Rosewood Drive, Danvers, MA 01923, 978-750-8400. For works that are not available on CCC please contact mpkbookspermissions@tandf.co.uk

Trademark notice: Product or corporate names may be trademarks or registered trademarks, and are used only for identification and explanation without intent to infringe.

Library of Congress Cataloging-in-Publication Data

Names: Selvamuthukumaran, M., editor.
Title: Applications of membrane technology for food processing industries /
    edited by M. Selvamuthukumaran.
Description: First edition. | Boca Raton : CRC Press, 2020. | Includes
    bibliographical references and index.
Identifiers: LCCN 2020028961 (print) | LCCN 2020028962 (ebook) | ISBN
    9780367226916 (hbk) | ISBN 9780429276408 (ebk)
Subjects: LCSH: Membrane filters--Industrial applications. | Processed
    foods. | Food industry and trade.
Classification: LCC TP159.M4 A66 2020 (print) | LCC TP159.M4 (ebook) |
    DDC 664--dc23
LC record available at https://lccn.loc.gov/2020028961
LC ebook record available at https://lccn.loc.gov/2020028962

ISBN: 978-0-367-22691-6 (hbk)
ISBN: 978-0-429-27640-8 (ebk)

Typeset in Palatino
by Deanta Global Publishing Services, Chennai, India

DEDICATION

I profoundly thank

the Almighty

My Family

My Friends

and

Everybody

*Who have constantly encouraged and helped
me to complete this book successfully*

Selvamuthukumaran. M

CONTENTS

Preface	ix
About the Editor	xi
Contributors	xiii

1 Introduction to Membrane Processing   1

   Carole C. Tranchant and M. Selvamuthukumaran

2 Frequently Used Membrane Processing Techniques for Food Manufacturing Industries   45

   Ulaş Baysan, Necmiye Öznur Coşkun, Feyza Elmas, and Mehmet Koç

3 Theoretical Approach behind Membrane Processing Techniques   97

   Komal Parmar

4 Deacidification of Fruit Juices by Electrodialysis Techniques   119

   M. Selvamuthukumaran

5 Clarification of Fruit Juices and Wine Using Membrane Processing Techniques   129

   Ismail Tontul

6 Microfiltration Techniques: Introduction, Engineering Aspects, Maintenance, and Its Application in Dairy Industries   155

   M. Selvamuthukumaran

7 Applications of Membrane Technology in Whey Processing   167

   Kirty Pant, Mamta Thakur and Vikas Nanda

CONTENTS

8 Membrane Technology for Degumming, Dewaxing and Decolorization of Crude Oil 201

M. Selvamuthukumaran

9 Retention of Antioxidants by Using Novel Membrane Processing Technique 211

Rahul Shukla, Mayank Handa, and Aakriti Sethi

10 Application of Membrane Processing Techniques in Wastewater Treatment for Food Industry 229

Rahul Shukla, Farhan Mazahir, Divya Chaturvedi and Vidushi Agarwal

Index 255

PREFACE

Membrane processing techniques help to separate chemical components based on molecular size under a specific pressure. The greatest advantage of this technique is that it is a non-thermal processing technique, which can retain enormous bioactive constituents to a greater extent. It is a less energy-intensive process and is widely used in the food processing industries, such as for clarification of fruit juices and wine, the concentration of milk, preparation of whey protein concentrate, water and waste treatment, etc.

This technique is also widely used for performing deacidification, demineralization, desalinization, purification processes in the food industries. It also reduces the microbial load in several liquid products, especially milk. This book will introduce membrane processing techniques in principle, theory and operation for achieving an efficient quality product. It describes the different types of membrane processing techniques *viz.* reverse osmosis, nanofiltration, ultrafiltration, electrodialysis, microfiltration, and pervaporation, along with applications, advantages and disadvantages.

It can be a ready reckoner for the food industry for manufacturing deacidified clarified fruit juices and wine by using an integrated membrane technique approach. It deals with the retention of antioxidants using novel membrane processing techniques. It deals with the application of membrane processing techniques in whey processing, which includes lactose and salt removal by ultrafiltration processes and its advantages. It explains the method for the degumming, dewaxing and decolorization of edible crude oils. The application of membrane processing techniques in wastewater treatment is also narrated for its efficient use.

This book will address professors, scientists, research scholars, students and industrial personnel for the clarification and concentration of liquid foods by using novel membrane processing techniques. It will narrate the optimal operating conditions for the production of concentrated products. This book will emphasize a novel approach to enhance antioxidants in end products by using integrated membrane processing techniques. This book will address the rectification problems for various types of fouling, which may occur during the membrane processing of foods. Readers will come to know about the current trends in the use of

PREFACE

membrane processing techniques in their application in the food processing industries. In a nutshell, this book will benefit food scientists, academicians, students and food industry personnel by providing in-depth knowledge about the membrane processing of foods for quality retention and also for efficient consumer acceptability.

I would like to express my sincere thanks to all the contributors; without their continuous support, this book would not have seen daylight. We would also like to express our gratitude to Mr. Steve Zollo and all other CRC Press staff, who have made every continuous cooperative effort to make this book a great standard publication at a global level.

M. Selvamuthukumaran

ABOUT THE EDITOR

Dr. M. Selvamuthukumaran is presently Associate Professor and Head of the Department of Food Technology, Hindustan Institute of Technology & Science, Chennai. He was a visiting Professor at Haramaya University, School of Food Science & Postharvest Technology, Institute of Technology, Dire Dawa, Ethiopia. He earned his PhD in Food Science from the Defence Food Research Laboratory affiliated with the University of Mysore, India. His core area of research is the processing of underutilized fruits for development of antioxidant-rich functional food products. He has contributed several technologies to Indian firms as an outcome of his research work. He has received several awards and citations for his research work. Dr. Selvamuthukumaran has published several international papers and book chapters in the area of antioxidants and functional foods. He has guided several national and international postgraduate students in the area of food science and technology.

CONTRIBUTORS

Vidushi Agarwal
Amity Institute of Biotechnology
Amity University
Noida, India

Ulas Baysan
Aydın Adnan Menderes University
Faculty of Engineering
Department of Food Engineering
Aydın, Turkey

Divya Chaturvedi
Amity Institute of Biotechnology
Amity University
Gurugram, India

Feyza Elmas
Aydın Adnan Menderes University
Faculty of Engineering
Department of Food Engineering
Aydın, Turkey

Mayank Handa
Department of Pharmaceutics
National Institute of
   Pharmaceutical Education and
   Research
Raebareli, India

Mehmet Koc
Aydın Adnan Menderes University
Faculty of Engineering
Department of Food Engineering
Aydın, Turkey

Farhan Mazahir
Department of Pharmaceutics
National Institute of
   Pharmaceutical Education and
   Research, Raebareli
Raebareli, India

Vikas Nanda
Department of Food Engineering
   and Technology
Sant Longowal Institute of
   Engineering and Technology
Longowal, India

Necmiye Öznur Coşkun
Aydın Adnan Menderes University
Faculty of Engineering
Department of Food Engineering
Aydın, Turkey

Kirty Pant
Department of Food Engineering
   and Technology
Sant Longowal Institute of
   Engineering and Technology
Longowal, India

Komal Parmar
ROFEL Shri G.M. Bilakhia College
   of Pharmacy
Vapi, India

CONTRIBUTORS

Aakriti Sethi
Department of Chemistry and
  Applied Biosciences
ETH Zurich
Zurich, Switzerland

Rahul Shukla
Department of Pharmaceutics
National Institute of
  Pharmaceutical Education and
  Research
Raebareli, India

Mamta Thakur
Department of Food Engineering
  and Technology
Sant Longowal Institute of
  Engineering and Technology
Longowal, India

Ismail Tontul
Necmettin Erbakan University
Faculty of Engineering and
  Architecture
Department of Food Engineering
Konya, Turkey

Carole C. Tranchant
School of Food Science and
  Nutrition
Faculty of Health Sciences and
  Community Services
Université de Moncton
New Brunswick, Canada

1

Introduction to Membrane Processing

Carole C. Tranchant and M. Selvamuthukumaran

Contents

Abbreviations		2
1.1	Introduction	2
1.2	Historical Overview of Membrane Technology	3
1.3	Operating Principle of Membrane Separation	5
1.4	Advantages and Limitations of Membrane Processing	8
	1.4.1 Advantages	9
	1.4.2 Limitations	11
1.5	Commercial Applications of Membrane Processing	12
1.6	Classes of Membranes	13
	1.6.1 Symmetric (or Isotropic) Membranes	18
	1.6.1.1 Microporous Membranes	18
	1.6.1.2 Non-Porous Dense Membranes	20
	1.6.1.3 Electrically Charged Membranes	20
	1.6.2 Asymmetric (or Anisotropic) Membranes	22
	1.6.3 Liquid Membranes	22
	1.6.4 Membrane Materials	23
	1.6.5 Membrane Configurations	24
1.7	Classification and Overview of Membrane Processes	27

APPLICATIONS OF MEMBRANE TECHNOLOGY

1.7.1	Pressure-Driven Membrane Processes		27
	1.7.1.1	Microfiltration and Ultrafiltration	27
	1.7.1.2	Nanofiltration and Reverse Osmosis	29
1.7.2	Concentration-Driven Membrane Processes		30
	1.7.2.1	Forward Osmosis	31
	1.7.2.2	Diffusion Dialysis and Dialysis	31
	1.7.2.3	Gas Separation	32
	1.7.2.4	Pervaporation	33
1.7.3	Electrically-Driven Membrane Processes		34
	1.7.3.1	Electrodialysis and Electroosmosis	34
1.7.4	Thermally-Driven Membrane Processes		35
	1.7.4.1	Membrane Distillation	35
1.8 Conclusion			36
References			37

ABBREVIATIONS

CA	cellulose acetate
ED	electrodialysis
FO	forward osmosis
GS	gas separation
LM	liquid membrane
MBR	membrane bioreactor
MD	membrane distillation
MF	microfiltration
NF	nanofiltration
PV	pervaporation
RO	reverse osmosis
UF	ultrafiltration

1.1 INTRODUCTION

Membrane processing is a technology of choice for separating and concentrating the components of a liquid or gaseous mixture according to their molecular size, shape or other relevant physicochemical properties. Membrane processing encompasses a variety of different processes depending on membrane characteristics and on the driving force of the permeate flow across the membrane. Aside from advantages such as efficiency and energy economy, their compact modular design and

INTRODUCTION TO MEMBRANE PROCESSING

operational simplicity enable continuous operation as well as wide-ranging applications (Ambrosi et al., 2017; Zhou and Husson, 2018). Membrane-based processes are generally considered to be a green technology as they operate without the addition of additives and chemicals, typically without heating (Dewettinck and Trung Le, 2011; Macedonio and Drioli, 2017). They can be used to process bio-based heat-sensitive materials and recover high-value bioactive and functional compounds such as bioactive peptides, polyphenols, prebiotics as well as flavour-active compounds (Akin et al., 2012; Saffarionpour and Ottens, 2018). Their applications in the agri-food and health sectors have expanded substantially during the past two decades as membrane-based operations are becoming increasingly competitive and economical compared with traditional concentration and separation technologies such as evaporation and freeze concentration.

The following sections provide an overview of membrane processing technology, its historical developments, principles of operation, advantages, limitations, applications in the food industries, classification of membranes and membrane processes, with consideration of the recent advances supporting the development of novel and high-quality foods and ingredients, including functional foods and nutraceuticals.

1.2 HISTORICAL OVERVIEW OF MEMBRANE TECHNOLOGY

Membrane technology has evolved into a mature technology and a major unit operation due to the discoveries of numerous researchers going back to the 18th century. A few of these pioneering discoveries are highlighted here. In 1748, a French clergyman and physicist named Nollet was the first to coin the term 'osmosis' to describe this natural process. Using a pig's bladder as a natural semipermeable membrane, he showed that solvent molecules from a water solution of low solute concentration could flow through the membrane into a solution of higher solute concentration made of alcohol (Strathmann, 2011). A few years later, Dutrochet constructed an osmometer for measuring the osmotic pressure and pointed to this pressure as the possible cause of the transport of water in plants. Cellulose nitrate, also known as nitrocellulose, the first semisynthetic polymer, was studied by Schoenbein in 1846. In the 1850s, Graham studied the diffusion of gases and liquids through various media. He studied *in vivo* dialysis and achieved the separation of colloids based on their molecular weight and concentration; he is also credited with coining the term 'dialysis' in 1861. In 1855, Fick used cellulose nitrate membranes to study diffusion and established the laws of diffusion, famously known as Fick's laws (Uragami, 2017). In 1867, Traube was the first to produce artificial semipermeable membranes made up of copper ferrocyanide precipitates, which laid the foundation for further research into osmotic pressure (Strathmann, 2011). Building on Traube's work in the 1870s, Pfeffer developed a thicker and more resistant membrane that could withstand greater pressure (Hendricks, 2006).

Another breakthrough came in 1907 when Bechhold devised a technique for preparing membranes of graded pore size, which opened the way for producing high-quality nitrocellulose-based membranes. These microporous membranes were commercialized in the 1930s and subsequently applied to microfiltration (Uragami, 2017). Bechhold is also credited with coining the term 'ultrafiltration' in 1910 (Hendricks, 2006). Microfiltration became more widely used in the 1950s with water treatment becoming its first prominent use for producing potable water. In 1959, Loeb and Sourirajan developed an asymmetric reverse osmosis membrane from cellulose acetate (CA), which rejected salt and totally dissolved solids, while allowing water to permeate at high fluxes at moderate operating pressures. This was a major advance in the production of potable water from seawater by reverse osmosis (Uragami, 2017).

In the 1960s, other membrane processing techniques, such as gas separation and membrane distillation, emerged. The first commercial application of membranes for gas applications in the 1980s was made feasible by the seminal work of Henis and Tripodi (Hagg, 2015). In the late 1980s, several membrane processing techniques, including microfiltration (MF), ultrafiltration (UF), reverse osmosis (RO) and electrodialysis (ED), emerged on a commercial scale with an increasing number of food applications. These advanced processes are widely used nowadays in the food and nutraceutical industries to separate and add value to a diverse range of food materials and byproducts. Nanofiltration (NF) and pervaporation (PV), which are relatively recent developments, are attracting increasing attention in these fields (Mohammad et al., 2019; Saffarionpour and Ottens, 2018), while membrane bioreactor (MBR) technology is in the emerging development phase with relatively few commercial applications at present (Mazzei et al., 2017).

1.3 OPERATING PRINCIPLE OF MEMBRANE SEPARATION

Membrane separation is based on the selective transport of certain substances through a semipermeable membrane. As shown schematically in Figure 1.1, this is achieved by interposing the membrane between a feed stream and a transfer or permeate stream, and by establishing conditions that provide a driving force for the transport of the solvent (generally water) and select solutes across the membrane from the feed to the permeate stream. The membrane is housed in a specific device, the membrane module, and acts as a barrier, separating two phases and restricting the transport of certain components in a selective manner. The rate of transport through the membrane depends on the driving force as well as on resistance, which depends on the membrane's properties and on the conditions in the fluid streams on each side of the membrane (Karel et al., 1995). Mass transport mechanisms involved in membrane separation processes include diffusion, solution-diffusion and convection (Buonomenna et al., 2012), while separation principles include adsorption, sieving (also known as size exclusion) and electrostatic phenomena (Padaki et al., 2015), as illustrated in Figure 1.2. Obtaining high-quality products relies on the use of semipermeable membranes, i.e., membranes with less resistance to the transport of solvent and select solutes than to the transport of the

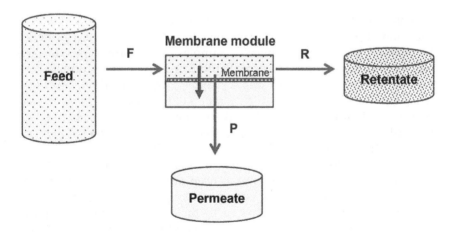

Figure 1.1 Schematic representation of membrane separation processes. F, R and P correspond respectively to the feed, retentate and permeate streams, while the small arrow inside the membrane module represents the transmembrane driving force.

APPLICATIONS OF MEMBRANE TECHNOLOGY

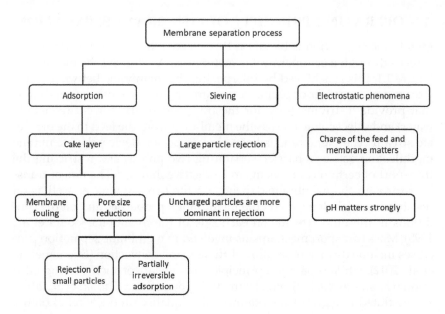

Figure 1.2 Separation principles involved in membrane processing. (Adapted from Padaki et al., 2015.)

components which are to be retained. Depending on the application, the value-added products derived from membrane processing may be either the retentate and/or the permeate. Examples of the former include various concentrates (e.g., diafiltered milk produced by UF, whey protein concentrates, concentrated fruit juices and coffee extracts), clarified beverages (e.g., juices, beer and wine), while examples of the latter include extended shelf-life (ESL) milk produced by MF, fresh and pure water obtained by RO, dealcoholized beer and wine, as well as volatile aroma compounds recovered from coffee, wine, beer and juices by PV.

The driving forces of membrane processes are gradients, i.e., differences in either hydraulic pressure, concentration, electrical potential or temperature. The different processes resulting from the application of these driving forces are summarized in Table 1.1. Their energy requirements are influenced by the driving force and the operating conditions, among other factors (Table 1.2). For instance, they tend to be higher in RO compared with NF and MF (Mulder, 1994), while in ED desalination they are proportional to the salinity level (Moran, 2018; Turek, 2002). Still, membrane-based processes generally require less energy than

INTRODUCTION TO MEMBRANE PROCESSING

Table 1.1 Classification of Membrane Separation Processes According to Their Driving Force[1]

Driving force	Pressure difference[2]	Concentration difference	Electrical potential difference	Temperature difference
Process	Microfiltration Ultrafiltration Nanofiltration Reverse osmosis	Forward osmosis Diffusion dialysis and dialysis Gas separation Pervaporation Liquid membrane extraction	Electrodialysis Electroosmosis	Membrane distillation

[1] Adapted from Buonomenna et al. (2012). Most of the processes are for liquid-phase separations, except for gas separation, pervaporation and membrane distillation which involve gas or a phase change from liquid to gas.

[2] Hydraulic pressure.

thermally-based separation systems such as evaporation (Mulder, 1994). In addition, developments in hybrid processing and energy recovery in membrane processes open the path towards higher energy efficiency (Gar, 2019; Macedonio and Drioli, 2017; Roy and Singha, 2017). Mass transport through membranes can be described by various mathematical relations. Most of these are semi-empirical and postulate models, such as Fick's law, Hagen-Poiseuille's law and Ohm's law. More comprehensive descriptions of transport processes, which are independent of the membrane structure and thus applicable to any membrane, have also been proposed (Giorno et al., 2015; Strathmann, 2001).

The main factors that influence the overall performance of membrane processes are summarized in Table 1.2. Membrane fouling, i.e., the deterioration in membrane performance due to the undesirable accumulation of organic and inorganic substances and particulate matter on the membrane surface and into its pores, is an important challenge inherent in all membrane separation processes. Its adverse effects range from permeate flux decline over filtration time, increased cleaning frequency and operational costs to reduced membrane life (Amy, 2008). Innovation in membrane and process design is driven to a large extent by the need to mitigate these effects, as discussed in numerous reports (e.g., Jepsen et al., 2018; Williams and Wakeman, 2000). Further details on the theory of the membrane separation of liquids and gases, including mass transfer mechanisms, transport

APPLICATIONS OF MEMBRANE TECHNOLOGY

Table 1.2 Main Factors Influencing the Performance of Membrane Processes

Factors	Contributing phenomena and influential process parameters (control strategies)
Energy consumption	• Driving force of the process • Membrane module design and membrane configuration • Operating conditions, e.g., pressure, temperature, feed concentration and flow rate, permeate flux • Membrane fouling and concentration polarization
Membrane fouling	• Adsorption of natural organic matter (NOM) • Adhesion of biological particulate matter, including cells and cell debris • Precipitation of inorganic compounds *Control strategies* • Flow configuration, e.g., cross-flow (tangential) filtration versus dead-end filtration • Membrane characteristics, e.g., organic/inorganic composition, structure, pore size, porosity, hydrophilicity/hydrophobicity, membrane configuration • Operating conditions, e.g., pressure, temperature, feed composition, concentration and flow rate, fluid velocity
Concentration polarization (precursor of fouling)	• Increased concentration of feed components near the surface of the membrane compared with their bulk concentration *Control strategies* • Same as above

resistances and the interrelated dynamics of membrane fouling and mass transport, can be found in Abdelrasoul et al. (2015), Roy and Singha (2017) and Strathmann (2011).

1.4 ADVANTAGES AND LIMITATIONS OF MEMBRANE PROCESSING

The advantages of membrane processing are the unique characteristics and the high-quality resulting products that it produces, as well as the process itself as compared with other concentration and separation techniques, as summarized below. These advantages contribute to the competitiveness of membrane technologies in the food and pharmaceutical sectors, among other fields.

1.4.1 Advantages

- Membrane processing enables the separation and concentration of a wide variety of components from liquid or gaseous mixtures, for instance, metal ions, salts, macromolecules and cells, according to their size, shape and sometimes electrical charge. The size of these components ranges from less than a nanometre to several micrometres (Figure 1.3). For gas mixtures, gas separation occurs according to their solubility in the membrane material.
- Various feedstocks including solutions, suspensions, emulsions as well as gaseous mixtures can be processed successfully using membrane technology.
- The quality of the resulting product is enhanced by membrane processing, either because specific components are concentrated (e.g., caseins, whey proteins), fractionated (e.g., antioxidants from

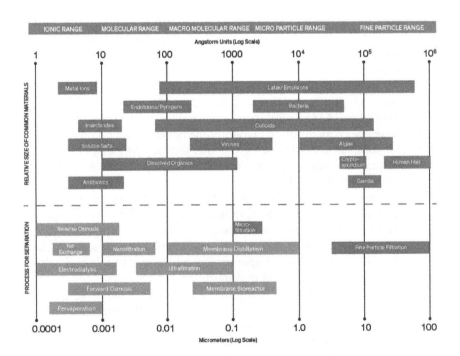

Figure 1.3 Solute and particle size range (upper panel) in relation to various membrane separation processes (lower panel). (Adapted from Warsinger et al., 2016.)

grape pomace, whey proteins), separated, recovered or purified (e.g., lactic acid, minerals, amino acids, peptides, proteins, antioxidants, prebiotics, volatile flavour compounds), removed when considered undesirable (e.g., microorganisms, pollutants, minerals, alcohol or compounds associated with haze and sediment formation in fruit juices, beers and wines), or produced or extracted through membrane biocatalysis (e.g., protein hydrolysates, thaumatin, carotenoids, prebiotics).

- Consistent and superior product quality can be obtained following process optimization.
- Membrane processes typically operate without the addition of additives or chemicals.
- They generally operate at relatively low temperatures (e.g., ambient or slightly higher or lower temperatures, depending on the application and process) (Kotsanopoulos and Arvanitoyannis, 2015). Heat-sensitive materials and their high-value bioactive constituents can be successfully processed without undesirable biochemical and bioactivity changes or losses of volatile components. When membrane filtration aims to reduce the microbial load of the product, the process is sometimes referred to as 'cold sterilization' or 'sterilization filtration' (e.g., MF of milk, fruit juices, wine, beer and pharmaceuticals) (American Membrane Technology Association [AMTA], 2014).
- The energy consumption of membrane operations is relatively low.
- The compact modular design of membrane processes and their operational simplicity enable continuous operation as well as scaling up to meet different needs.
- The membranes are environmentally friendly in general and increasingly efficient. Membranes with high selectivity and permeability enable large volumes to be processed. They generate relatively little waste during their operation. They are safe, stable under the recommended operating conditions, reusable and relatively easy to operate and maintain.
- Membrane technology is highly versatile. The choice of membrane, filtration configuration, number of filtration modules and operating conditions can be adapted as needed for specific purposes. Moreover, membrane operations can be used as integrated/hybrid systems, in combination with other processes or as a combination of different types of membranes in one process (Buonomenna et al., 2012; Gebreeyessus, 2019; Mikhaylin

INTRODUCTION TO MEMBRANE PROCESSING

et al., 2016; Salgado et al., 2017). Such configurations facilitate the recovery of energy, which lowers energy consumption (Gar, 2019; Macedonio and Drioli, 2017). Recent developments in membrane bioreactor (MBR) technology enable the production of high-value bioactive and functional compounds through biocatalytic processes (Mazzei et al., 2017).

- Investment and operation costs are relatively low. Membrane operations generally require fewer operators and less operator attention as compared with conventional concentration and separation processes. Equipment operation and cleaning can be automated. Some membrane systems are designed to clean the membranes in place (cleaning-in-place), without removing them from the system.
- Membrane processing is generally considered to be a green technology. Membrane operations are one of the most attractive candidates for meeting the requirements of sustainability by decreasing the energy consumption, production costs, equipment size as well as waste generation (Buonomenna et al., 2012; Macedonio and Drioli, 2017).

1.4.2 Limitations

The main limitations of membrane processing are as follows:

- Membrane fouling is a major challenge for the efficient operation of membrane systems. It causes a decline in the membrane permeate flux and quality over the filtration time, leading to higher energy use and decreased filtration performance. Common types of fouling include organic and biological fouling, colloidal fouling, scaling and metal oxide fouling (Wang et al., 2014; Warsinger et al., 2016). Various strategies can be implemented to retard fouling and decrease its severity, including feed pre-treatment, choice of membrane material and morphology, filtration in the crossflow configuration, and the use of high shear, depending on the specific process and feed material considered. This topic has been thoroughly discussed in the recent literature (Jepsen et al., 2018; Nguyen et al., 2012; Saleh and Gupta, 2016; Sanaei and Cummings, 2017; Williams and Wakeman, 2000).
- Concentration polarization, which is intrinsically associated with the selective character of membranes, also causes the permeate

APPLICATIONS OF MEMBRANE TECHNOLOGY

flux to decline over time. It refers to the reversible accumulation of rejected material close to the membrane surface and is often a precursor of membrane fouling. It is generally not considered fouling as it is more readily reversible than fouling. As for fouling, various methods are available to control concentration polarization (Giacobbo et al., 2018).

- The relatively low chemical and thermal stability of some membranes (e.g., some polymeric membranes) and the inherent trade-off relation between permeability and selectivity limit the applicability of certain membranes. Membranes with low selectivity decrease the extent of concentration that can be achieved in the retentate, while the permeate contains a significant amount of the components which are intended to be retained.
- Membrane elements require cleaning on a regular basis, when the decline in permeate flux becomes significant. They also need to be replaced after some time, for instance, after one to three years depending on membrane type (e.g., organic or polymeric versus inorganic), composition and usage, among other factors. Some membranes have shorter lifespans than others.
- The disposal of certain filtration byproducts (e.g., retentates with high concentrations of salt, microorganisms or other contaminants) may be a problem in some situations, especially when it is not feasible to discharge them into a sanitary sewer system for treatment at a wastewater facility.

1.5 COMMERCIAL APPLICATIONS OF MEMBRANE PROCESSING

Since its first large-scale applications for water treatment in the 1950s, membrane technology has found widespread and growing use in diverse industrial sectors, for instance, in the chemical, food, biotechnology and pharmaceutical industries, as well as in the environmental engineering field which supports numerous industrial activities. Moreover, membrane technology is considered a promising alternative for biofuel production (Kumar et al., 2019; Shuit et al., 2012).

In the food industry, membrane technology is a major tool for improving the quality, safety and shelf-life of food and beverage products. It was adopted at a relatively early stage by the dairy industry in the late 1960s and has subsequently gained acceptance in other branches (Cuperus, 1998;

Mohammad et al., 2019). Today, it serves a broad range of applications in the agri-food sector, as illustrated in Table 1.3. Details are presented in Section 1.7 and in the following chapters of this book. These applications fall into two main categories, i.e., food processing applications and post-processing applications. The first category spans a variety of food and beverage products, as well as more recent applications in functional foods and nutraceuticals (Akin et al., 2012; Kotsanopoulos and Arvanitoyannis 2015; Nazir et al., 2019) which involve the recovery, separation and purification of naturally occurring compounds capable of exerting beneficial biological activities *in vivo* (e.g., antioxidant, antihypertensive or prebiotic activity). These applications often entail novelty and innovation based on membrane processes (Bazinet and Firdaous, 2011; Daufin et al., 2001), as in the example of extended shelf-life milk produced using the Bactocatch® MF process (Elwell and Barbano, 2006). Post-processing applications include wastewater treatment from food processing and, more recently, the recovery of valuable compounds from underutilized food processing byproducts and wastes for purposes in various sectors such as the food, chemical, biotechnology, pharmaceutical and cosmetic industries.

Commercial applications of membrane technology outside the realm of food processing are similarly diverse (AMTA, 2014). They include but are not limited to the separation and purification of bacterial vaccines, antibiotics and enzymes by MF or UF (Hadidi et al., 2015), the removal of arsenic from water by NF and RO (Nicomel et al., 2016), the desalination of seawater and brackish water by ED or RO, the recovery of nickel and sulphuric acid from ED waste by dialysis (Tongwen and Weihua, 2004), organic solvent recovery by PV, as well as gas separation (e.g., methane from carbon dioxide (CO_2) and other gases present in biogas, and helium from natural gas) (Bernardo and Clarizia, 2013).

1.6 CLASSES OF MEMBRANES

Membranes used in separation, concentration and purification processes today are highly diverse in composition, structure and functionality. Most of them are synthetic or semisynthetic and may be classified in various ways according to their constituent materials, structure and morphology, configuration, separation mechanism, separation process as well as preparation methods (Saleh and Gupta, 2016). Based on materials, structure and morphology, membranes can be organic (e.g., polymeric) or inorganic (e.g., ceramics), solid or liquid, symmetric or asymmetric, homogeneous

APPLICATIONS OF MEMBRANE TECHNOLOGY

Table 1.3 Representative Applications of Membrane Processing Technologies in the Agri-Food Industries[1]

Process	Applications	References
Microfiltration	• Removal of bacteria and other microorganisms from milk, beverages and honey • Separation and fractionation of milk fat globules • Fractionation of milk proteins, production of casein concentrates	Dhineshkumar and Ramasamy (2017), Kotsanopoulos and Arvanitoyannis (2015), Subramanian et al. (2007)
	• Clarification and stabilization of fruit juices, tomato juice, beer and wine	Ambrosi et al. (2014), Kotsanopoulos and Arvanitoyannis (2015)
	• Dewaxing and decolorization of vegetable oils	Manjula and Subramanian (2006)
	• Treatment of food processing effluents in the dairy, seafood and other sectors	Kotsanopoulos and Arvanitoyannis (2015)
Ultrafiltration	• Fractionation of whey proteins and milk fat • Concentration of milk proteins, lactose and calcium • Clarification and stabilization of fruit juices, clarification of honey, sugarcane juice and tea extracts	Dhineshkumar and Ramasamy (2017), Kotsanopoulos and Arvanitoyannis (2015), Wang et al. (2011)
	• Degumming of vegetable oils	Manjula and Subramanian (2006)
	• Production of enzyme enriched honey	Subramanian et al. (2007)
	• Separation and fractionation of valuable components such as proteins, phospholipids, galacto-oligosaccharide prebiotics, bioactive peptides and polyphenols, and their recovery from food processing effluents	Afonso and Borquez (2002), Akin et al. (2012), Cassano et al. (2018), Kotsanopoulos and Arvanitoyannis (2015), Nazir et al. (2019)

(Continued)

INTRODUCTION TO MEMBRANE PROCESSING

Table 1.3 (Continued) Representative Applications of Membrane Processing Technologies in the Agri-Food Industries[1]

Process	Applications	References
Nanofiltration	• Partial demineralization of whey • Concentration of whey proteins • Removal of lactic acid from whey, removal of lactose from milk	Dhineshkumar and Ramasamy (2017), Kotsanopoulos and Arvanitoyannis (2015), Mohammad et al. (2019)
	• Sugar level control in beverages and grape must • Reduction of the alcohol content of wines	Salgado et al. (2017)
	• Fractionation of antioxidants from grape pomace • Desalination of soy sauce	Kotsanopoulos and Arvanitoyannis (2015)
	• Degumming of vegetable oils	Manjula and Subramanian (2006)
	• Wastewater treatment and recovery of bioactive compounds from food processing effluents in the dairy, brewing, wine and other sectors	Akin et al. (2012), Kotsanopoulos and Arvanitoyannis (2015), Nazir et al. (2019)
Reverse osmosis	• Concentration of whey and milk • Demineralization of whey • Concentration of fruit juices, egg white, coffee, sugar solutions, syrups, natural extracts and flavours	Dhineshkumar and Ramasamy (2017), Kotsanopoulos and Arvanitoyannis (2015)
	• Clarification and dealcoholization of beer and wine	Ambrosi et al. (2014)
	• Water demineralization, purification, desalination and recovery of bioactive compounds from food processing effluents	Akin et al. (2012), Kotsanopoulos and Arvanitoyannis (2015), Nazir et al. (2019)

(Continued)

15

APPLICATIONS OF MEMBRANE TECHNOLOGY

Table 1.3 (Continued) Representative Applications of Membrane Processing Technologies in the Agri-Food Industries[1]

Process	Applications	References
Forward osmosis	• Concentration of fruit and vegetable juices, orange peel press liquor, sugar and protein solutions, natural food colourants and bioactive extracts • Concentration of brine in water desalination processes	Rastogi (2016), Terefe et al. (2016)
Diffusion dialysis and dialysis	• Recovery of acids and alkalis from waste streams, removal/recovery of organic acids generated in processes such as fermentation	Luo et al. (2011)
	• Dealcoholization of beer	Ambrosi et al. (2014)
Gas separation	• Production of oxygen-enriched air • Separation of methane and CO_2 from biogas	Bernardo and Clarizia (2013)
Pervaporation	• Recovery of natural volatile aroma compounds from fruit juices, beer and wine	Ambrosi et al. (2014), Salgado et al. (2017)
	• Dealcoholization of beer	Castro-Munoz (2019)
	• Recovery of ethanol from fermentation broths	Peng et al. (2011)
	• Solvent dehydration and solvent removal from wastewater	Roy and Singha (2017), Wang et al. (2011)
Liquid membrane extraction	• Separation of amino acids, fatty acids, lactic acid and inorganic salts • Recovery/removal of organic solvents and hazardous substances such as heavy metals from waste streams	Demirel and Gerbaud (2019), Kumar et al. (2018)

(Continued)

INTRODUCTION TO MEMBRANE PROCESSING

Table 1.3 (Continued) Representative Applications of Membrane Processing Technologies in the Agri-Food Industries[1]

Process	Applications	References
Electrodialysis	• Demineralization of whey, skim milk, beet and cane molasses and distillery vinasse • Desalination of water, mussel cooking juice, fish sauce and soy sauce, concentration of brine	Chehayeb et al. (2017), Kotsanopoulos and Arvanitoyannis (2015), Mikhaylin and Bazinet (2016)
	• Removal of potassium hydrogen tartrate for wine tartaric stabilization • Deacidification of fruit juices and acidity control of other liquid food products • Production of enriched protein fractions, fractionation of amino acids, bioactive peptides and antioxidants • Recovery and concentration of lactic acid and other organic acids	Kotsanopoulos and Arvanitoyannis (2015), Mikhaylin and Bazinet (2016)
	• Wastewater treatment and recovery of bioactive compounds from food processing effluents	Akin et al. (2012), Mikhaylin and Bazinet (2016), Nazir et al. (2019)
Membrane distillation	• Concentration of fruit juices, skim milk and whey	Hausmann et al. (2014)
	• Flavour recovery • Dealcoholization of beer and wine and alcohol removal from fermentation broths	Saffarionpour and Ottens (2018)
	• Wastewater treatment and recovery of valuable compounds from the liquid effluents from vegetable oil and meat processing	El-Abbassi et al. (2013), Mostafa et al. (2017)

[1] Details on the operating conditions and types of membranes preferred in each application can be found in the references provided in this table.

17

or heterogeneous, porous or non-porous, electrically charged or neutral, as summarized in Figure 1.4. The main classes of membranes are briefly reviewed here based on structure and morphology, materials and configurations.

1.6.1 Symmetric (or Isotropic) Membranes

Symmetric membranes are also referred to as isotropic membranes. They are characterized by a uniform chemical composition and physical structure over their cross-section. Their resistance to mass transfer is determined by their total thickness (Strathmann, 2001). Symmetric membranes are often subdivided into microporous membranes, non-porous dense membranes and electrically charged membranes (Lee et al., 2016), as illustrated in Figure 1.5 (upper panel).

1.6.1.1 Microporous Membranes

Porous or microporous membranes have pores that are much larger than the molecular size of the permeating substances (permeants). They are also referred to as millipore or ultrafiltration membranes. These membranes are similar to conventional filters, except for their smaller pore size (diameter), typically from 10^{-2} to 10^1 μm, which enables the separation of molecules and particles at the submicron scale (Lee et al., 2016). The separation of solutes by microporous membranes is based primarily on the

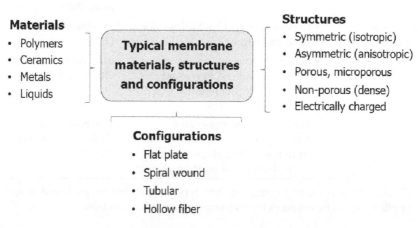

Figure 1.4 Typical materials, structures and configurations used in membrane processes.

INTRODUCTION TO MEMBRANE PROCESSING

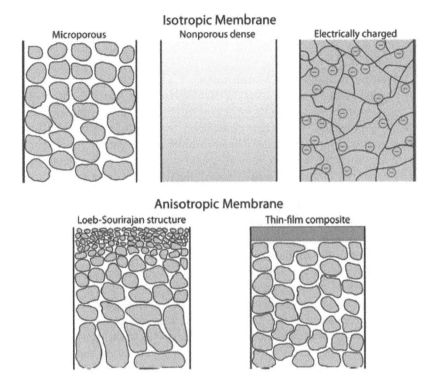

Figure 1.5 Different classes of membranes based on structure/morphology (Lee et al., 2016).

solute molecular size in relation to the pore size of the membrane and involves sieving (size exclusion) and adsorption mechanisms. Hydraulic flow of solvent and low molecular weight solutes occurs through these membranes, whereas solutes of high molecular weight (larger than the membrane pores) are retained at the surface. The main disadvantage of porous membranes is membrane fouling due to pore blocking, which increases the membrane resistance and leads to flux decline over time. As noted in Section 1.4.2, various strategies are available for controlling membrane fouling and improving membrane regeneration/fouling recovery. Microporous membranes can be prepared from different materials, mainly polymers, ceramics, metals and, more recently, graphene oxide (Lee et al., 2016). Microporous membranes are generally used in MF and UF for liquid separations. By convention, MF and UF membranes have pore diameters in the range of 10^{-1}–1 μm and 10^{-2}–10^{-1} μm, respectively

(Figure 1.3 and Table 1.4). Their structures are considerably more porous in general when compared with RO and NF membranes (Warsinger et al., 2016).

1.6.1.2 Non-Porous Dense Membranes

Non-porous dense membranes consist of a dense film which enables the separation of components of a mixture (generally small molecules in the gas phase or in solution) based on their solubility and diffusivity in the membrane material. These membranes are also referred to as homogenous membranes. They are generally made of dense polymers and are used to separate smaller molecules than those separated using porous membranes. The separation of solutes in non-porous membranes is governed by their relative transport rates within the membrane, which are determined by their solubility and diffusivity in the membrane material. This occurs through the solution-diffusion mechanism, whereby the component with the highest solubility or diffusivity permeates more rapidly through the membrane (Lee et al., 2016). Components of similar size can be separated if their solubility and diffusivity in the membrane are different enough. Since the components need to dissolve into the membrane, the properties of the dense polymeric material are important for the separation process (Werber et al., 2016). The main disadvantage of dense membranes is their relatively low flux. Therefore, the dense film is usually made extremely thin (e.g., 0.1–1 μm) and is deposited on top of an asymmetric membrane as its 'skin' layer (selective barrier) supported by a highly porous substructure. Non-porous dense membranes are commonly used in RO, NF as well as PV and gas separation (Shon et al., 2013; Synder Filtration, 2019).

1.6.1.3 Electrically Charged Membranes

Electrically charged membranes consist of gel-type polymer structures carrying fixed charges. They are also called ion-exchange membranes as they attract oppositely charged ions (i.e., counter-ions) and allow their transport through the membrane (Strathmann, 2011). Co-ions (i.e., mobile ions of the same charge as the fixed ions), meanwhile, are excluded from the membrane due to Coulomb repulsion. Charged membranes include monopolar, bipolar and mosaic membranes (Ji et al., 2017). Monopolar membranes carrying positive charges are called anion-exchange membranes, while those carrying negative charges are referred to as cation-exchange membranes. A combination of these membranes as adjacent layers results in a bipolar membrane. Other membranes, known as

INTRODUCTION TO MEMBRANE PROCESSING

Table 1.4 Overview of Pressure-Driven Membrane Processes Used for Liquid-Phase Separation and Their Main Characteristics[1,2]

Process	Commonly used membranes	Pore size (µm) (nm)	Operating pressure (bar)	Pure water flux (L/m²/h)	Components retained (examples)
Microfiltration (MF)	Porous symmetric	10^{-1}–1 100–1,000	0.5–5	500–10,000	Suspended particles, cells, bacteria, yeasts, moulds, fat globules
Ultrafiltration (UF)	Porous asymmetric	10^{-2}–10^{-1} 10–100	1–10	100–2,000	Colloids, macromolecules (e.g., proteins, enzymes, polysaccharides), viruses
Nanofiltration (NF)	Finely porous asymmetric/ composite	10^{-3}–10^{-2} 1–10	10–40	20–200	Divalent ions (soluble salts), small organic molecules (e.g., lactose and other sugars, organic acids, amino acids, small peptides, natural pigments, phenolic compounds, alcohol, natural organic matter, pesticides)
Reverse osmosis (RO)	Non-porous asymmetric/ composite[3]	10^{-4}–10^{-3} 0.1–1	20–100	10–100	Monovalent ions (dissolved salts), metal ions, nitrates, small molecules (e.g., sugars, organic acids, amino acids, alcohol, organic pollutants)

[1] Adapted from Nicomel et al. (2016) and Shon et al. (2013).
[2] Typical characteristics and orders of magnitude are presented in this table.
[3] RO membranes are so dense (pore size < 10^{-3} µm or 1 nm) that they are generally considered as non-porous.

charged mosaic membranes, contain channels of both positive and negative charges. This property enables the concurrent transport of anions and cations, resulting in the enrichment of solute (e.g., salts) concentration in the permeate (Summe et al., 2018). Charged membranes generally have a microporous structure. Thus, separation is achieved both by charge exclusion and size exclusion. It is also influenced by the concentration of the salts or charged organic molecules in the solution (Lee et al., 2016). Charged membranes are extensively used in ED. Other applications include NF and, more recently, hybrid processes such as ED with NF, in which one of the conventional ED ion-exchange membranes is replaced by a charged NF membrane in an ED cell (Bazinet and Moalic, 2011).

1.6.2 Asymmetric (or Anisotropic) Membranes

Asymmetric membranes, also called anisotropic membranes, consist of several layers, each with a different structure and permeability (Figure 1.5, lower panel). They include two main types, namely, Loeb-Sourirajan membranes and composite membranes, such as thin-film membranes and solution-coated membranes (Lee et al., 2016). Loeb-Sourirajan membranes are homogenous in composition, but their pore size and porosity vary in the different layers of the membrane. The denser layer is on the surface, while the underlying sublayer is increasingly porous. Thin-film membranes in comparison consist of two distinct substructures made of different materials. The surface layer is a very thin, dense layer, called the 'skin' or active layer (typically 0.1–1 µm in thickness), made of a highly cross-linked polymer which acts as the selective barrier. It is deposited on a thicker (100–200 µm) and highly porous substructure which provides mechanical support. Thin-film composite membranes are widely used in RO, NF, gas separation and PV. The separation characteristics and permeation rates of asymmetric membranes are determined by the properties of the surface layer, which in turn depend on the nature of the material and resulting structure. Solution-diffusion and charge exclusion play a major role in the separation process. The resistance to mass transfer is determined mainly by the thickness of the surface layer, while the porous support merely affects the separation (Shon et al., 2013; Lee et al., 2016).

1.6.3 Liquid Membranes

Liquid membranes (LM) consist of a thin liquid film in which a selective carrier, solubilized in the membrane solvent, enables the transport of certain components at a relatively high rate across the membrane. This type of transport is known as facilitated transport and occurs by solution-diffusion due to a concentration difference (Krull et al., 2008; Li et al., 2015). Various components of liquid or gas mixtures can be efficiently separated by this mechanism, including heavy metal ions, amino acids, fatty acids, lactic acid, water and inorganic salts (Demirel and Gerbaud, 2019; Kumar et al., 2018). Specific molecular recognition (i.e., high membrane selectivity) can be achieved with the use of proper carriers. However, maintaining the integrity of the liquid film and its properties during the separation process can be a challenge. LM function either in a supported or unsupported form, respectively called 'supported LM' and 'emulsion LM.' In the supported form, the pores of a porous membrane are filled with the selective liquid material (solvent and carrier). In the emulsion form, the LM is a thin oil film stabilized by surfactants in a double emulsion and the inner droplets of the double emulsion act as the receiving phase (Raghuraman et al., 1994). LM are a promising technique for the separation and recovery of organic solvents, hazardous compounds as well as valuable organic components from the waste streams of various industries. Other promising applications include the separation and recovery of CO_2 and methane from industrial gas streams (Hojniak et al., 2013).

1.6.4 Membrane Materials

Filtration membranes can be made of different types of materials, classified as organic, inorganic and composite materials, using various fabrication methods. This research area, and particularly the development of novel membrane materials, is an important factor driving improvements in membrane performance and process efficiency. The choice of material is dictated mainly by the desired intrinsic characteristics of the membrane (e.g., pore size, selectivity and permeability), its resistance to fouling, mechanical strength, thermal and chemical stability (e.g., resistance to disinfectants, solvents and high/low pH), cost as well as durability (Warsinger et al., 2016). Materials for solid membranes are the focus of the remainder of this section. For LM, details can be found in the reviews of Krull et al. (2008) and Parhi (2013).

Synthetic organic polymers are most commonly used across the different types of membranes in current commercial applications, including MF, UF, NF, RO, PV and gas separation (Baker and Low, 2014; Lee et al., 2016; Nicomel et al., 2016; Peng et al., 2011). MF and UF membranes are often made of the same polymeric materials, typically polysulfone, polyvinylidene

APPLICATIONS OF MEMBRANE TECHNOLOGY

fluoride, polyacrylonitrile and copolymers of polyacrylonitrile and polyvinyl chloride. Other frequently used polymers include CA-cellulose nitrate blends, nylons and polytetrafluoroethylene (MF membranes) as well as polyethersulfone and polysulfone amide (UF) (Lee et al., 2016). NF membranes, which exhibit intermediate properties between those of UF and RO membranes, generally consist of a modified form of UF membranes (e.g., sulfonated polysulfone) or are prepared from the same materials as RO membranes, typically polyamide composite membranes (polyamide thin-film coated onto a porous polysulfone support), CA or blends of CA and cellulose triacetate (Lee et al., 2016). The main drawback of most polymeric membranes is their relatively low resistance to chemically aggressive environments (e.g., chlorine, oxidants, extreme pH conditions) and high temperatures, which often results in a shorter lifetime of the membrane and higher operation cost (Warsinger et al., 2016).

Inorganic membranes have received increasing attention over the past two decades due to their higher thermal, chemical and mechanical stability, ease of cleaning in applications with high fouling conditions, and superior durability compared with current polymeric membranes. Higher cost, however, has hindered their widespread use. In recent years, hybrid (inorganic-organic) materials have emerged as a promising strategy to overcome the limitations associated with either polymeric or inorganic membranes. Inorganic materials used in membrane manufacture include porous ceramics, also known as metal oxides (e.g., Al_2O_3, TiO_2, ZrO_2, ZnO and SiO_2), carbon-based materials (e.g., graphene and graphene oxides), metals (e.g., Ag, Cu, Pd), composites containing two materials or more (e.g., TiO_2-SiO_2, TiO_2-ZrO_2 and Al_2O_3-SiC) and nanoparticle composites (e.g., Ag-TiO_2, Zn-CeO_2 and zeolites) (Ersan et al., 2017; Lee et al., 2016). Inorganic membranes may be used in gas separation, PV, as well as in liquid separation as a membrane support or when high temperatures are required. Ceramic membranes, for instance, may be used in MF and UF (Lee et al., 2016), while Pd-based metallic membranes can be used to produce hydrogen from methane in catalytic membrane reactors (Kian et al., 2018; Palo et al., 2018).

1.6.5 Membrane Configurations

Membrane configuration refers to the geometry of the membrane and its position in relation to the flow of the feed and of the permeate. It also determines the way in which the membrane is packed inside the membrane module (Berk, 2018). Four main types of membrane configurations

INTRODUCTION TO MEMBRANE PROCESSING

are commonly used in commercial applications such as food processing, namely, flat plate, spiral wound, tubular and hollow fibre configurations (Figures 1.4 and 1.6). The membrane geometry is planar in the first two and cylindrical in the two others. Other (novel) configurations have been proposed (see, for instance, Jie et al., 2012 and Park et al., 2015), but have not yet reached commercial status on a large scale.

In the flat plate configuration, also known as plate-and-frame or flat sheet (Figure 1.6a), the membranes are arranged in vertical or

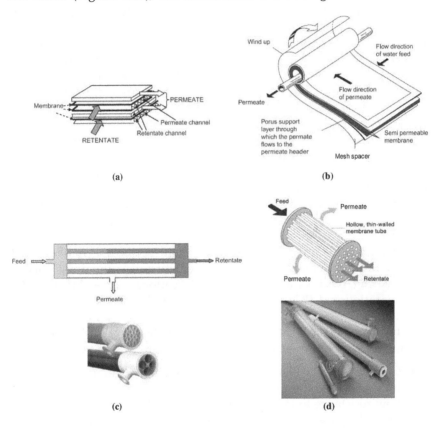

Figure 1.6 Commercially available membrane modules: (a) flat plate, (b) spiral wound, (c) tubular and (d) hollow fibre modules. (Adapted from Berk, 2018; and Lutz, 2015.)

25

APPLICATIONS OF MEMBRANE TECHNOLOGY

horizontal stacks. Flat plate modules are generally limited to MF, UF and LM as they cannot withstand very high pressures; they also have relatively high costs and low packing densities (low surface-area-to-volume ratio). They are particularly suitable in applications in which the feed stream contains high amounts of foulants and/or has a high viscosity (Berk, 2018; Warsinger et al., 2016). Spiral wound modules (Figure 1.6b) consist of flat sheet membranes, spacers and porous support layer tightly wrapped around a central porous tube of small diameter which collects the permeate. This configuration results in high packing densities and is easy to manufacture, but it is highly sensitive to fouling. It is commonly used in NF and RO processes as well as in UF and MF, preferably with feeds that have low concentrations of total suspended solids (TSS) (Warsinger et al., 2016). Tubular modules (Figure 1.6c) are made of several tubes, each with a diameter in the range of 10–30 mm, installed as parallel bundles inside a shell. The possibility of maintaining a high tangential velocity in the feed stream, which promotes turbulence and minimizes fouling, makes the tubular configuration particularly suitable for feeds with high TSS concentrations or that must be strongly concentrated. Tubular membranes are also easy to clean. Their packing density, however, is relatively low (Berk, 2018). Hollow fibre membranes (Figure 1.6d) consist of several thousand very thin tubes with a small hollow portion of 0.5–1.25 mm in diameter. These fibres are much thinner than the tubes in the tubular setup. Hollow fibre membranes are an attractive alternative to flat membranes due to their high packing density and to the possibility of backwashing, but they are limited to applications such as MF, UF, LM, gas separation and PV in which the pressure does not exceed 2–4 bar (Krull et al., 2008; Warsinger et al., 2016).

The main desirable characteristics guiding the selection of membrane configuration are summarized below (Berk, 2018):

- Compactness, that is, the ability to pack a high membrane surface into a module of limited volume.
- Low resistance to tangential flow (i.e., low friction, low-pressure drop along the flow channels and low-energy expenditure).
- A high degree of turbulence at the retentate side to minimize fouling and enhance mass transfer.
- Uniform velocity distribution and no regions of stagnant flow.
- Easy cleaning, disinfection and maintenance.
- Low cost per unit of membrane area.

Flow configuration is another important consideration for membrane modules. It refers to either cross-flow or dead-end filtration. In cross-flow filtration (or tangential flow filtration), the fluid flow is applied tangentially across the surface of the membrane, while in dead-end filtration, the flow is applied perpendicular to the membrane surface. The cross-flow configuration is preferred in most applications as it creates conditions such as shear and turbulent flow which reduce fouling. It also increases membrane lifespan by helping prevent irreversible fouling. Dead-end filtration is limited to applications in which the TSS load of the feed is very low.

1.7 CLASSIFICATION AND OVERVIEW OF MEMBRANE PROCESSES

Membrane operation processes are often classified according to their driving force into pressure-driven, concentration-driven, electrically-driven and thermally-driven processes (Table 1.1). The fundamentals of the technologies included in each category are summarized below, together with representative examples of their applications in food processing.

1.7.1 Pressure-Driven Membrane Processes

Pressure-driven membrane processes (MF, UF, NF and RO) work by applying pressure to the feed solution. The transmembrane hydraulic pressure gradient serves as the driving force inducing the flow of solvent (water in most applications) and some solutes through the semipermeable membrane. The four technologies included in this category are used for liquid-phase separation and can be distinguished by their decreasing membrane pore size, i.e., MF > UF > NF > RO, and by other characteristics summarized in Table 1.4. Processes with the smallest pore sizes (NF and RO) require a higher driving force, i.e., higher operating pressure (≥ 10 bar), and are thus classified as high-pressure processes, while MF and UF are considered low-pressure processes.

1.7.1.1 Microfiltration and Ultrafiltration

MF and UF are similar in that their driving force is a relatively low pressure applied to the feed liquid. Microporous membranes are used in both processes but with different pore sizes (approximately 10^{-1}–1 µm and

10^{-2}–10^{-1} μm, respectively). Consequently, MF is mainly used to separate suspended solids or particles, bacteria, yeasts, moulds, cells and some colloids such as fat globules, whereas UF separates smaller colloids and macromolecules (e.g., polysaccharides and proteins) and viruses (Table 1.4). In both processes, the separation occurs mainly through sieving. Since the first commercial-scale applications of MF in water and wastewater treatment operations and in the dairy industry (e.g., production of casein concentrates, removal of milk fat globules and debacterization of milk to produce ESL milk), this technology has found several applications in food sectors as diverse as the fruit and tomato juice, brewing, winemaking, vegetable oil and seafood industries, among others (Dhineshkumar and Ramasamy, 2017; Kotsanopoulos and Arvanitoyannis, 2015) (Table 1.3). Fractionation, clarification, reduction of the microbial content, removal of suspended particles, emulsion separation and biomass separation from fermented broths are some of the main targets of MF processes in these applications.

Likewise, UF is well established in water and wastewater treatment as well as in the dairy industry to concentrate milk prior to the manufacture of dairy products, to concentrate individual milk components (e.g., whey, lactose and calcium) and to fractionate milk fat and whey proteins. Other sectors such as the beverage industries and, to a lesser extent, the honey, cereal and meat processing industries, have also adopted UF (Kotsanopoulos and Arvanitoyannis, 2015), followed by the functional food and nutraceutical sectors in recent years (Akin et al., 2012; Loginov et al., 2013; Nazir et al., 2019). In a special type of UF, known as diafiltration, the retentate is diluted with water and then reprocessed by UF to further reduce the concentration of soluble components in the permeate and increase the concentration of the retentate (Dhineshkumar and Ramasamy, 2017). Diafiltration is mainly applied in the dairy sector to produce milk protein concentrates, also known as 'diafiltered milk,' and whey protein concentrates. Tight UF membranes have recently been developed to overcome the limitations of conventional UF membranes. Due to their relatively low molecular weight cut-off (MWCO) (e.g., 1–3 kDa) compared with traditional UF membranes (50–100 kDa), tight UF membranes are highly effective to concentrate low molecular weight compounds such as peptides and polyphenols. The recovery of bioactive polyphenols from agri-food byproducts is among their promising applications (Cassano et al., 2018).

Other promising approaches for enhancing the efficiency of UF and MF separation include integrated/hybrid processes in which MF and UF

INTRODUCTION TO MEMBRANE PROCESSING

membranes are used in combination in the same process, or in combination with another membrane or non-membrane separation technology (Mikhaylin et al., 2016). In wastewater treatment, the combination of MF and UF membranes in a membrane bioreactor (MBR) is relatively common. In this configuration, the membranes are submerged inside the bioreactor and vacuum can be used to separate the treated water from the retained solids (Warsinger et al., 2016). The emerging food applications of MF and UF membranes in MBR and biocatalytic membrane reactors have been reviewed by Mazzei et al. (2017).

1.7.1.2 Nanofiltration and Reverse Osmosis

NF and RO are similar in that they are designed to remove dissolved organic and non-organic components including salts. They are also known as 'loose reverse osmosis' and 'hyperfiltration,' respectively. Both technologies require high hydraulic pressures and use similar membrane materials much less porous than in MF and UF. NF removes many of the same solutes as RO, but to a lesser degree due to different membrane characteristics (Warsinger et al., 2016) (Table 1.4). NF membranes generally consist of thin-film composite membranes with a dense surface layer whose pore size and MWCO (approximately 10^{-3}–10^{-2} μm and 180–2,000 Da, respectively) allow the passage of some neutral solutes and monovalent ions, while multivalent ions and small organic molecules are retained. In addition to molecular weight, membrane polarity and the affinity of the solutes for membrane material play an important role in NF and RO.

A distinct advantage of NF membranes over RO membranes is their higher water permeability and the fact that they can be operated at lower pressures, thus reducing the energy consumption (Warsinger et al., 2016). Many commercial applications of NF are geared toward water softening and salt fractionation for compounds of low molecular weight, including organic salts. NF is frequently applied in desalination operations, separation of minerals from wastewater, removal of contaminants from liquid streams, wastewater treatment for water recovery, and recovery of high-value organic compounds (e.g., peptides and enzymes). NF is well established in the dairy industry where representative applications include whey demineralization, whey protein concentration and lactic acid recovery. In the wine industry, NF is successfully applied to increase the sugar level in grape must, to lower the alcohol content and to fractionate valuable antioxidants present in grape pomace (Kotsanopoulos and Arvanitoyannis, 2015; Salgado et al., 2017) (Table 1.3). NF applications extend to other sectors such as sugar, edible oil and seafood processing, soy sauce desalination as well as the separation of bioactive compounds for use as nutraceuticals or in functional foods (Akin et al., 2012).

RO is characterized by the application of very high pressures, which exceed the osmotic pressure of the feed solution, and by the sub-nanometre pore size of the membrane (Table 1.4). Due to their low MWCO (10–100 Da), RO membranes have very high rejection rates for various organic and non-organic solutes, up to 95–99% depending on the compounds, resulting in the production of water of high purity (Kotsanopoulos and Arvanitoyannis, 2015). RO retains most dissolved salts as well as organic molecules that commonly contaminate wastewater, including contaminants of emerging concern (CEC) such as sex steroid hormones, microplastics and new-generation pesticides (Novotna et al., 2019; Romeyn et al., 2016; Warsinger et al., 2016). RO is also effective at removing larger particulates (e.g., dissolved natural organic matter) and pathogenic microorganisms if these are not already removed by upstream pre-treatment membranes (e.g., MF or UF) (Warsinger et al., 2016). Water demineralization and purification, desalination and water recovery from waste streams are among the main applications of RO. Its wastewater treatment applications are widespread in the dairy, meat, fish and cereal processing industries. Some of the largest commercial food applications of RO include the concentration of whey from cheese manufacture and whey demineralization (Table 1.3). RO is also applied to concentrate and purify fruit juices, tomato juice, citric acid, enzymes, polyphenols, fermentation liquors and vegetable oils; to concentrate egg white, milk, coffee, sugar solutions, syrups, natural extracts and flavours; to clarify wine and beer and dealcoholize these beverages (Akin et al., 2012; Ambrosi et al., 2014; Koseoglu et al., 1991; Kotsanopoulos and Arvanitoyannis, 2015).

1.7.2 Concentration-Driven Membrane Processes

Technologies in this category include forward osmosis, diffusion dialysis, dialysis, gas separation and pervaporation. Separation processes based on LM (see Section 1.6.3), which are also referred to as 'liquid membrane extraction,' may also be included in this category. Depending on the technology, several terms are used in the literature to describe their driving force, for instance, the gradient in osmotic pressure, partial pressure, vapour pressure, concentration or chemical potential. As all these terms can be related to the solute concentration, these technologies are commonly classified as concentration-driven membrane processes.

INTRODUCTION TO MEMBRANE PROCESSING

1.7.2.1 Forward Osmosis

Forward osmosis (FO) is an emerging technology enabling the concentration of liquid products through the removal of water across a semipermeable dense (non-porous) hydrophilic membrane that separates two aqueous solutions, i.e., a feed solution and a draw solution, having different osmotic pressures. FO relies on the generation of a large osmotic pressure differential across the membrane, which acts as a driving force of the flow of pure water from the dilute feed solution to the concentrated draw solution (Rastogi, 2016). This process is also known as 'direct osmosis,' 'engineered osmosis' or 'manipulated osmosis' in order to differentiate it from RO and the naturally occurring osmosis process in which no draw solutions are used. In contrast to FO, RO uses hydraulic pressure as the driving force for separation, which serves to counteract the osmotic pressure gradient that would otherwise favour the flux of water from the permeate to the feed (Nazir et al., 2019; Rastogi, 2016). The draw solutions in FO are commonly prepared from sodium chloride (NaCl). Other osmotic agents such as $CaCl_2$, $MgCl_2$ and $MgSO_4$ have also been tested (Rastogi, 2016). The draw solute used to prepare this solution must not pass through the membrane and should have a higher osmotic pressure than the feed solution that is being concentrated.

FO is gaining importance for the concentration of liquid products or streams such as liquid foods, natural colourants, wastewater streams as well as seawater and brackish water (Rastagi, 2016). In addition to being an economically feasible alternative to evaporation for the concentration of liquid foods, FO has several advantages over pressure-driven membrane processes that make it commercially attractive. These include the potential for energy saving as FO operates at ambient pressure and temperature, a lower membrane fouling propensity as the solids are not compressed against the membrane during the operation, and the possibility to achieve high concentration factors (Terefe et al., 2016). Concentration polarization remains an important issue in FO and warrants additional research on the design of dense membranes with improved performance (Eyvaz et al., 2018). Promising food applications of FO include the concentration of fruit and vegetable juices, orange peel press liquor, protein solutions, natural food colourants and bioactive extracts such as anthocyanin and betalain extracts (Rastagi, 2016) (Table 1.3).

1.7.2.2 Diffusion Dialysis and Dialysis

Diffusion dialysis is an ion-exchange membrane separation process driven by a concentration gradient. During this process, the solutes carrying a negative or positive charge are transported across the semipermeable membrane from the compartment with the highest solute concentration to the compartment with the lowest concentration. The membrane, an anion-exchange or cation-exchange membrane, is permeable only to anions or cations, respectively, and is impermeable to the solvent (typically water), as described in Section 1.6.1.3. Although diffusion dialysis is similar in principle to dialysis, its main distinguishing feature is the use of ion-exchange membranes. The main applications of diffusion dialysis are the separation and recovery of acids and alkalis from waste solutions, which can be performed in a cost-effective and environmentally friendly way (Luo et al., 2011).

Diffusion dialysis is successfully applied in the metallurgical industries, the steel industry among others, while its applications in the bio-industries include the removal and recovery of organic acids used or generated in some food and pharmaceutical processes. Fermentation processes, for instance, generate different carboxylic acids and their salts (carboxylates) which can be effectively separated by diffusion dialysis (Luo et al., 2011). The main limitation of diffusion dialysis is that the concentration of the recovered solution is limited by the equilibrium. For this reason, diffusion dialysis is not as predominant as some other membrane technologies such as RO and electrodialysis. However, its unique low-energy and low-pollution characteristics may make it a more attractive technology in the future (Luo et al., 2011). Dialysis is similar in principle to diffusion dialysis except that it does not use ion-exchange membranes. Its applications in the food industry include the production of low-alcohol beers (Ambrosi et al., 2014).

1.7.2.3 Gas Separation

In contrast to the previous membrane processes, which are used for liquid-phase separation, membrane gas separation (GS) enables the separation of one or several gases from a gas mixture. Both the feed and the permeate are in a gaseous state and the driving force inducing the mass transport through the semipermeable membrane is the difference in partial pressure often generated by applying a vacuum at the permeate side or by applying pressure to the feed stream. The gas molecules are separated based on differences in solubility and diffusivity in the membrane (Bernardo and Clarizia, 2013). The GS membranes in current commercial use are mostly asymmetric polymeric membranes with a dense surface layer (Baker and Low, 2014). The literature on GS indicates that LM may also be used but on a relatively small scale (Krull et al., 2008). The main

INTRODUCTION TO MEMBRANE PROCESSING

industrial applications of GS include the removal/recovery of hydrogen, CO_2 and organic vapours from gas streams, natural gas dehydration, air dehumidification and the production of oxygen-enriched air (Bernardo and Clarizia, 2013). Oxygen-enriched air has diverse chemical, medical and industrial applications. In food packaging, for instance, high-oxygen packaging may be used to help preserve the red colour of red meats. Promising developments include the integration of GS membrane units in hybrid systems (Bernardo and Clarizia, 2013).

1.7.2.4 Pervaporation

In contrast to gas separation, the feed in pervaporation (PV) is a liquid. This liquid mixture contacts one side of the membrane, while the permeate is removed as a vapour at the other side. Transport through the selective membrane is induced by a difference in vapour pressure between the feed solution and permeate vapour (Ambrosi et al., 2014). PV is based on the vapour/liquid equilibrium, as is membrane distillation (MD). In most configurations, the vapour pressure difference in PV is generated by applying a vacuum at the permeate side, whereas the upstream membrane side is at ambient pressure. The permeate vapour is subsequently condensed by cooling. PV membranes are generally asymmetric membranes with a dense surface layer made of hydrophilic or hydrophobic polymers (Roy and Singha, 2017). Hydrophilic membranes permit water permeation, while hydrophobic membranes allow the permeation of organic compounds. Separation results from the different permeation rates of the components through the membrane.

PV is an attractive technology for systems that are difficult to separate by other conventional or membrane-based techniques, for instance, azeotropic mixtures, heat-sensitive materials and organic compounds highly diluted in complex aqueous solutions. PV was found to be effective for recovering natural volatile aroma compounds from fruit juices, beer and wine, for beer dealcoholization (Ambrosi et al., 2014; Castro-Munoz, 2019; Salgado et al., 2017) and for removing dilute volatile organic compounds from wastewater (Roy and Singha, 2017; Wang et al., 2011) (Table 1.3). Other applications include the recovery of ethanol from fermentation broths (Peng et al., 2011), solvent dehydration and solvent removal from industrial wastewater. Promising approaches for enhancing the efficiency of the separation include integrated/hybrid processes in which PV is combined with another technique such as NF (Salgado et al., 2017) or MD (Kotsanopoulos and Arvanitoyannis, 2015).

1.7.3 Electrically-Driven Membrane Processes

1.7.3.1 Electrodialysis and Electroosmosis

Electrodialysis (ED) is an electrochemical process in which ionized species such as salts are removed from solution by applying an electric field across an ion-selective membrane. In ED, the transmembrane transport of ions, also known as electromigration, is driven by the difference in electrical potential (i.e., voltage) across the membrane. The selectivity of the membrane is determined by the ionic groups chemically bound to its structure, for instance, ammonium (NH_4^+) and sulfonate (SO_4^{2-}) groups which are used for anion-exchange and cation-exchange membranes, respectively. Alternate stacking of the two types of membranes in a multi-membrane ED cell allows for the simultaneous removal of anions and cations from the feed based on the dilution-concentration principle (Bazinet, 2004; Nazir et al., 2019). Water is also transported across the membrane as the ions migrating through the membrane are surrounded by a hydration shell. This phenomenon, known as electroosmosis, can be used for removing water in operations such as dewatering and concentration (Menon et al., 2019).

The pore size of commercial ED membranes generally varies between 10^{-3} and 10^{-2} µm. Monopolar polymeric membranes are the most common form used in ED processes at present, while ED with bipolar membranes is a growing field of interest. Bipolar membranes carry positive charges on one side of the membrane, negative charges on the other side and both sides are joined by a hydrophilic junction. They dissociate water molecules through electrolysis and produce a flow of hydroxyl ions and protons through their anion- and cation-exchange interfaces, respectively (Bazinet, 2004). Bipolar membranes are mostly used for the production of acids and bases, and in processes that require acidification or alkalization to separate specific compounds (Campione et al., 2018). In addition, hybrid processes, namely, ED with MF, UF or NF, were recently developed to separate charged organic molecules, including polyphenols and bioactive peptides, based on molecular charge and size. They consist of a multi-membrane ED cell in which some of the ED monopolar membranes are replaced with a MF, UF or NF membrane (Bazinet and Moalic, 2011).

ED is a technology of choice for the concentration and separation of ions and charged molecules from aqueous solutions. Commercial applications of ED with monopolar membranes include desalination (seawater, brackish water and industrial wastewater), water demineralization, as well as the separation and concentration of various substances such as

INTRODUCTION TO MEMBRANE PROCESSING

organic acids (e.g., lactic acid and citric acid) and ammonium sulphate in the biochemical, biotechnological and pharmaceutical industries (Moresi and Sappino, 1998; Warsinger et al., 2016; Wee et al., 2005). ED was recently found to be an economically promising technology for the recovery of nitrogen from wastewater (Ward et al., 2018). In the food industry, ED with monopolar membranes has found numerous applications in the processing of dairy products, wines and fruit juices, as well as in the treatment of food processing effluents and byproducts, for instance, demineralization of distillery vinasse, desalination of mussel cooking juice and recovery of bioactive peptides from crab byproducts (Kotsanopoulos and Arvanitoyannis, 2015; Nazir et al., 2019) (Table 1.3). Applications in the dairy industry include the demineralization of whey and skim milk. In winemaking, ED is applied to stabilize wines by removing potassium bitartrate (Fidaleo and Moresi, 2006; Mikhaylin and Bazinet, 2016).

Hybrid processes involving ED with MF were found to be advantageous to solve the problems of microbiological stability, clarification, tartrate stabilization and oxidation of wine in one step (Daufin et al., 2001). Likewise, processes combining ED and UF were found to be efficient for the separation of bioactive peptides (Pouliot et al., 2006) and chitosan oligomers (Aider et al., 2008). Promising applications of ED with bipolar membranes in food processing include the production of caseinates, the fractionation of whey proteins, the production of lactic acid from whey fermentation (Bazinet, 2004), the separation of amino acids from fermentation broths (Buchbender and Wiese, 2018), the deacidification of fruit juice (Vera et al., 2009) and the inhibition of enzymatic browning in cloudy apple juice (Lam Quoc et al., 2000). One of the benefits of bipolar membranes is that they can be used to modulate the pH of food products without the addition of acid or base, which alleviates the undesirable effects of dilution (Bazinet and Doyen, 2017).

1.7.4 Thermally-Driven Membrane Processes

1.7.4.1 Membrane Distillation

Membrane distillation (MD) is an emerging technology based on the evaporation of a volatile solvent such as water or alcohol through a hydrophobic membrane permeable only to the gas molecules (e.g., water vapour). The driving force of this process, a temperature difference across the membrane, results in a difference in vapour pressure which induces the transport of the volatile components in the feed through the membrane and to the permeate (distillate) stream where they are condensed (Deshmukh et al., 2018). The membrane surface on the feed side is in direct contact with the liquid phase (hot side), while the permeate (cold side) may be a gas, an aqueous solution or vacuum, depending on the configuration. Evaporation takes place at the feed-membrane interface. Polymeric microporous membranes made of the same materials as MF membranes (e.g., polypropylene, polyvinylidene difluoride or polytetrafluoroethylene) are commonly used in MD because of their high hydrophobicity. Ceramic membranes are also gaining popularity (Ramlow et al., 2019). Pore sizes in MD range from 10^{-2} to 1 μm, depending on the application.

The driving force of MD and the use of hydrophobic membranes enable high retention of non-volatile components such as salts, dissolved organic molecules, macromolecules and colloidal particles. In fact, one of the main advantages of MD compared with other membrane processes is the possibility of producing highly concentrated retentates, while simultaneously producing water of high quality (Mostafa et al., 2017). Moreover, MD operates at lower pressures than pressure-driven membrane processes and at lower feed temperatures than conventional distillation (Ramlow et al., 2019). It is a promising technology for the desalination of seawater and brackish water, the treatment of wastewaters from various industries and the concentration of aqueous solutions (Warsinger et al., 2017). In the context of the food industry, MD has shown potential for concentrating fruit juices, skim milk and whey (Hausmann et al., 2014), dealcoholization of beer and wine, alcohol removal from fermentation broths, flavour recovery (Saffarionpour and Ottens, 2018) as well as wastewater treatment, including the recovery of valuable components from the liquid effluents from edible oil and meat processing (El-Abbassi et al., 2013; Mostafa et al., 2017) (Table 1.3).

1.8 CONCLUSION

Membrane processing is a versatile and rapidly evolving technology used to separate and concentrate the components of liquid or gaseous mixtures. It encompasses diverse processes characterized by distinct driving forces and membrane characteristics. The membrane plays a pivotal role governing the overall efficiency of these processes. Over the last 40 years, membrane-based processes have found a continuously increasing number of applications in the agri-food industries where they have become major tools for improving the quality, safety and shelf-life of food and beverage products. They also provide the industry with efficient tools for creating high added value by manufacturing novel products and by recovering valuable compounds such as bioactive molecules from traditionally underutilized byproducts or waste. Most, if not all, sectors of the food industry today are benefiting from some type of membrane-based processing, including the dairy, beverage, edible oil, egg, meat, seafood cereal and flavour sectors, as well as in the newer and fast-growing sectors of functional foods and nutraceuticals.

Membrane processes have crucial advantages over other technologies. They are generally considered to be a greener and economical technology as they require relatively low-energy inputs and capital costs and operate without the addition of chemicals, typically without heating or at lower temperatures than conventional technologies. In the food and beverage industry, membrane processes are vital drivers of product novelty and innovation, and of improvements in competitiveness. Moreover, they contribute to curbing the environmental impacts of food processing by reducing the energy consumption and enabling a more efficient and environmentally friendly treatment of post-processing effluents and wastewater. Further developments in integrated and hybrid membrane operations are expected to enhance the benefits of membrane processing in this industry due to the synergy effects than can be achieved by combining different technologies. Continuous innovations and improvements in membrane performances and operating conditions that minimize fouling and concentration polarization will undoubtedly support these developments and widen the application range of membrane processes. The significance of membrane processing in the food and beverage industry is further discussed in this book, with a focus on recent technological developments and new prospects in a wide range of sectors.

REFERENCES

Abdelrasoul, A., Doan, H., Lohi, A., and Cheng, C. H. 2015. Mass transfer mechanisms and transport resistances in membrane separation process. In: *Mass Transfer: Advancement in Process Modelling*, ed. M. Solecki, 15–40. London: IntechOpen.

Afonso, M. D., and Borquez, R. 2002. Review of the treatment of seafood processing wastewaters and recovery of proteins therein by membrane separation processes – Prospects of the ultrafiltration of wastewaters from the fish meal industry. *Desalination* 142(1):29–45.

Aider, M., Brunet, S., and Bazinet, L. 2008. Electroseparation of chitosan oligomers by electrodialysis with ultrafiltration membrane (EDUF) and impact on electrodialytic parameters. *Journal of Membrane Science* 309(1–2):222–32.

Akin, O., Temelli, F., and Koseoglu, S. 2012. Membrane applications in functional foods and nutraceuticals. *Critical Reviews in Food Science and Nutrition* 52(4):347–71.

Ambrosi, A., Cardozo, N. S., and Tessaro, I. C. 2014. Membrane separation processes for the beer industry: A review and state of the art. *Food and Bioprocess Technology* 7(4):921–36.

American Membrane Technology Association (AMTA) 2014. *Industrial Applications of Membranes*. Stuart: AMTA.

Amy, G. 2008. Fundamental understanding of organic matter fouling of membranes. *Desalination* 231(1–3):44–51.

Baker, R. W., and Low, B. T. 2014. Gas separation membrane materials: A perspective. *Macromolecules* 47(20):6999–7013.

Bazinet, L. 2004. Electrodialytic phenomena and their applications in the dairy industry: A review. *Critical Reviews in Food Science and Nutrition* 44(7–8):525–44.

Bazinet, L., and Doyen, A. 2017. Antioxidants, mechanisms, and recovery by membrane processes. *Critical Reviews in Food Science and Nutrition* 57(4):677–700.

Bazinet, L., and Firdaous, L. 2011. Recent patented applications of ion-exchange membranes in the agrifood sector. *Recent Patents on Chemical Engineering* 4:207–16.

Bazinet, L., and Moalic, M. 2011. Coupling of porous filtration and ion-exchange membranes in an electrodialysis stack and impact on cation selectivity: A novel approach for sea water demineralization and the production of physiological water. *Desalination* 1–3(1–3):356–63.

Berk, Z. 2018. *Food Process Engineering and Technology*. Cambridge: Academic Press.

Bernardo, P., and Clarizia, G. 2013. 30 years of membrane technology for gas separation. *Chemical Engineering Transactions* 32:1999–2004.

Buchbender, F., and Wiese, M. 2018. Efficient concentration of an amino acid using reactive extraction coupled with bipolar electrodialysis. *Chemical and Engineering and Technology* 41(12):2298–305.

Buonomenna, M. G., Golemme, G., and Perrotta, E. 2012. Membrane operations for industrial applications. In: *Advances in Chemical Engineering*, ed. Z. Nawaz, 543–62. Rijeka: InTech Europe.

Campione, A., Gurreri, L., Ciofalo, M., Micale, G., Tamburini, A., and Cipollina, A. 2018. Electrodialysis for water desalination: A critical assessment of recent developments on process fundamentals, models and applications. *Desalination* 434:121–60.

Cassano, A., Conidi, C., Ruby-Figueroa, R., and Castro-Munoz, R. 2018. Nanofiltration and tight ultrafiltration membranes for the recovery of polyphenols from agro-food by-products. *International Journal of Molecular Sciences* 19(2):351.

Castro-Munoz, R. 2019. Pervaporation-based membrane processes for the production of non-alcoholic beverages. *Journal of Food Science and Technology* 56(5):2333–44.

Chehayeb, K. M., Farhat, D. M., Nayar, K. G., and Lienhard, J. H. 2017. Optimal design and operation of electrodialysis for brackish-water desalination and for high-salinity brine concentration. *Desalination* 420:167–82.

Cuperus, F. P. 1998. Membrane processes in agro-food: State-of-the-art and new opportunities. *Separation and Purification Technology* 14(1–3):233–9.

Daufin, G., Escudier, J.-P., Carrère, H., Bérot, S., Fillaudeau, L., and Decloux, M. 2001. Recent and emerging applications of membrane processes in the food and dairy industry. *Transactions of the Institution of Chemical Engineers* 79:89–102.

Demirel, Y., and Gerbaud, V. 2019. *Nonequilibrium Thermodynamics: Transport and Rate Processes in Physical, Chemical and Biological Systems*. Amsterdam: Elsevier.

Deshmukh, A., Boo, C., Karanikola, V., Lin, S., Straub, A. P., Tong, T., Warsinger, D. M., and Elimelech, M. 2018. Membrane distillation at the water-energy nexus: Limits, opportunities, and challenges. *Energy and Environmental Science* 11(5):1177–96.

Dewettinck, K., and Trung Le, T. 2011. Membrane separations in food processing. In: *Alternatives to Conventional Food Processing*, ed. A. Proctor, 184–253. London: Royal Society for Chemistry.

Dhineshkumar, V., and Ramasamy, D. 2017. Review on membrane technology applications in food and dairy processing. *Journal of Applied Biotechnology and Bioengineering* 3:399–407.

El-Abbassi, A., Hafidi, A., Khayet, M., and García-Payo, M. C. 2013. Integrated direct contact membrane distillation for olive mill wastewater treatment. *Desalination* 323:31–8.

Elwell, M. W., and Barbano, D. M. 2006. Use of microfiltration to improve fluid milk quality. *Journal of Dairy Science* 89(E. Suppl.):E10–30.

Ersan, G., Apul, O. G., Perreault, F., and Karanfil, T. 2017. Adsorption of organic contaminants by graphene nanosheets: A review. *Water Research* 126:385–98.

Eyvaz, M., Arslan, S., Imer, D., Yuksel, E., and Koyuncu, I. 2018. Forward osmosis membranes – A review: Part I. In: *Osmotically Driven Membrane Processes – Approach, Development and Current Status*, eds. H. Du, A. Thompson, X. Wang, 11–40. London: IntechOpen.

Fidaleo, M., and Moresi, M. 2006. Electrodialysis applications in the food industry. *Advances in Food and Nutrition Research* 51:265–360.

Gar, M. C. 2019. Renewable energy-powered membrane technology: Cost analysis and energy consumption. In: *Current Trends and Future Developments on (Bio)-Membranes. Renewable Energy Integrated with Membrane Operations*, eds. A. Basil, A. Cassano, A. Figoli, 85–110. Amsterdam: Elsevier.

Gebreeyessus, G. D. 2019. Status of hybrid membrane–ion-exchange systems for desalination: A comprehensive review. *Applied Water Science* 9(5):135.

APPLICATIONS OF MEMBRANE TECHNOLOGY

Giacobbo, A., Bernardes, A. M., Filipe Rosa, M. J., and de Pinho, M. N. 2018. Concentration polarization in ultrafiltration/nanofiltration for the recovery of polyphenols from winery wastewaters. *Membranes* 8(3):46.

Giorno, L., Strathmann, H., and Drioli, E. 2015. Mathematical description of mass transport in membranes. In: *Encyclopedia of Membranes*, eds. E. Drioli, L. Giorno. Heidelberg: Springer-Verlag.

Hadidi, M., Buckley, J., and Zydney, A. L. 2015. Ultrafiltration behavior of bacterial polysaccharides used in vaccines. *Journal of Membrane Science* 490:294–300.

Hagg, M. B. 2015. Gas permeation: Permeability, permeance, and separation factor. In: *Encyclopedia of Membranes*, eds. E. Drioli, L. Giorno. Heidelberg: Springer-Verlag.

Hausmann, A., Sanciolo, P., Vasiljevic, T., Kulozik, U., and Duke, M. 2014. Performance assessment of membrane distillation for skim milk and whey processing. *Journal of Dairy Science* 97(1):56–71.

Hendricks, D. 2006. *Water Treatment Unit Processes: Physical and Chemical*. Boca Raton: CRC Press, Taylor & Francis.

Hojniak, S. D., Khan, A. L., Hollóczki, O., Kirchner, B., Vankelecom, I., Dehaen, W., and Binnemans, K. 2013. Separation of carbon dioxide from nitrogen or methane by supported ionic liquid membranes (SILMs): Influence of the cation charge of the ionic liquid. *Journal of Physical Chemistry B* 117(48):15131–40.

Jepsen, K. L., Bram, M., Pedersen, S., and Yang, Z. 2018. Membrane fouling for produced water treatment: A review study from a process control perspective. *Water* 10(7):847.

Ji, Y. L., Gu, B. X., An, Q., and Gao, C. G. 2017. Recent advances in the fabrication of membranes containing 'ion pairs' for nanofiltration processes. *Polymers* 9(12):715.

Jie, L., Liu, L., Yang, F., Liu, F., and Liu, Z. 2012. The configuration and application of helical membrane modules in MBR. *Journal of Membrane Science* 392–393:112–21.

Karel, M., Fennema, O. R., and Lund, D. B. 1995. *Physical Principles of Food Preservation*. New York: Marcel Dekker, Inc.

Kian, K., Woodall, C. M., Wilcox, J., and Liguori, S. 2018. Performance of Pd-based membranes and effects of various gas mixtures on H_2 permeation. *Environments* 5(12):128.

Koseoglu, S. S., Rhee, K. C., and Lucas, E. W. 1991. Membrane separations and applications in cereal processing. *Cereal Foods World* 377:376–83.

Kotsanopoulos, K., and Arvanitoyannis, I. S. 2015. Membrane processing technology in the food industry: Food processing, wastewater treatment, and effects on physical, microbiological, organoleptic, and nutritional properties of foods. *Critical Reviews in Food Science and Nutrition* 55(9):1147–75.

Krull, F. F., Fritzmann, C., and Melin, T. 2008. Liquid membranes for gas/vapor separations. *Journal of Membrane Science* 325(2):509–19.

Kumar, R., Ghosh, A. K., and Pal, P. 2019. Sustainable production of biofuels through membrane-integrated systems. *Separation and Purification Reviews* 1:1–22.

Kumar, A., Thakur, A., and Panesar, P. S. 2018. Lactic acid extraction using environmentally benign green emulsion ionic liquid membrane. *Journal of Cleaner Production* 181:574–83.

Lam Quoc, A., Lamarche, F., and Makhlouf, J. 2000. Acceleration of pH variation in cloudy apple juice using electrodialysis with bipolar membranes. *Journal of Agricultural and Food Chemistry* 48(6):2160–6.

Lee, A., Elam, J. W., and Darling, S. B. 2016. Membrane materials for water purification: Design, development, and application. *Environmental Science: Water Resource and Technology* 2(1):17–42.

Li, Y., Wang, S., He, G., Wu, H., Panab, F., and Jiang, Z. 2015. Facilitated transport of small molecules and ions for energy-efficient membranes. *Chemical Society Reviews* 44(1):103–18.

Loginov, M., Boussetta, N., Lebovka, N., and Vorobiev, E. 2013. Separation of polyphenols and proteins from flaxseed hull extracts by coagulation and ultrafiltration. *Journal of Membrane Science* 442:177–86.

Luo, J., Wu, C., Xu, T., and Wu, Y. 2011. Diffusion dialysis-concept, principle and applications. *Journal of Membrane Science* 366(1–2):1–16.

Lutz, H. 2015. Modules. In: *Ultrafiltration for Bioprocessing*, ed. H. Lutz. Cambridge: Woodhead Publishing.

Macedonio, F., and Drioli, E. 2017. Membrane engineering for green process engineering. *Engineering* 3(3):290–8.

Manjula, S., and Subramanian, S. 2006. Membrane technology in degumming, dewaxing, deacidifying, and decolorizing edible oils. *Critical Reviews in Food Science and Nutrition* 46(7):569–92.

Mazzei, R., Piacentini, E., Gebreyohannes, A. Y., and Giorno, L. 2017. Membrane bioreactors in food, pharmaceutical and biofuel applications: State of the art, progresses and perspectives. *Current Organic Chemistry* 21(17):1–31.

Menon, A., Mashyamombe, T., Kaygen, E., Nasiri, M., and Stojceska, V. 2019. Electro-osmosis dewatering as an energy efficient technique for drying food materials. *Energy Procedia* 161:123–32.

Mikhaylin, S., and Bazinet, L. 2016. Electrodialysis in food processing. In: *Reference Module in Food Science*, 1–6. Amsterdam: Elsevier.

Mikhaylin, S., Nikonenko, V., Pourcelly, G., and Bazinet, L. 2016. Hybrid bipolar membrane electrodialysis/ultrafiltration technology assisted by a pulsed electric field for casein production. *Green Chemistry* 18(1):307.

Mohammad, A. W., Teow, Y., Ho, K., and Rosnan, N. A. 2019. Recent developments in nanofiltration for food applications. In: *Nanomaterials for Food Applications*, eds. A. L. Rubio, M. J. Rovira, M. Sanz, L. Gómez-Mascaraque, 101–19. Amsterdam: Elsevier.

Moran, S. 2018. *An Applied Guide to Water and Effluent Treatment Plant Design*. Amsterdam: Elsevier.

Moresi, M., and Sappino, F. 1998. Economic feasibility study of citrate recovery by electrodialysis. *Journal of Food Engineering* 35(1):75–90.

Mostafa, M. G., Zhu, B., Cran, M., Dow, N., Milne, N., Desai, D., and Duke, M. 2017. Membrane distillation of meat industry effluent with hydrophilic polyurethane coated polytetrafluoroethylene membranes. *Membranes* 7(4):55.

Mulder, M. 1994. Energy requirements in membrane separation processes. In: *Membrane Processes in Separation and Purification*, eds. J. G. Crespo, K. W. Boddeker. Proceedings of the NATO Advanced Science Institute, 443–70. Dordrecht: Springer.

Nazir, A., Khan, K., Maan, A., Zia, R., Giorno, L., and Schroën, K. 2019. Membrane separation technology for the recovery of nutraceuticals from food industrial streams. *Trends in Food Science and Technology* 86:426–38.

Nguyen, T., Roddick, A., and Fan, L. 2012. Biofouling of water treatment membranes: A review of the underlying causes, monitoring techniques and control measures. *Membranes* 2(4):804–40.

Nicomel, N. R., Leus, K., Folens, K., Van Der Voort, P., and Du Laing, G. 2016. Technologies for arsenic removal from water: Current status and future perspectives. *International Journal of Environmental Research and Public Health* 13:62.

Novotna, K., Cermakova, L., Pivokonska, L., Cajthaml, T., and Pivokonsky, M. 2019. Microplastics in drinking water treatment – Current knowledge and research needs. *Science of the Total Environment* 667:730–40.

Padaki, M., Murali, R. S., Abdullaha, M. S., Misdana, N., Moslehyani, A., Kassim, M. A., Hilal, N., and Ismail, A. F. 2015. Membrane technology enhancement in oil-water separation. A review. *Desalination* 357:197–207.

Palo, E., Salladini, A., Morico, B., Palma, V., Ricca, A., and Iaquaniello, G. 2018. Application of Pd-based membrane reactors: An industrial perspective. *Membranes* 8(4):101.

Parhi, P. K. 2013. Supported liquid membrane principle and its practices: A short review. *Journal of Chemistry*: 618236.

Park, S. R., Kim, J. H., Ali, A., Macedonio, F., and Drioli, E. 2015. A novel approach to synthesize helix wave hollow fiber membranes for separation applications. *Journal of Membrane and Separation Technology* 4(1):8–14.

Peng, P., Shi, B., and Lan, Y. 2011. A review of membrane materials for ethanol recovery by pervaporation. *Separation Science and Technology* 46(2):234–46.

Pouliot, J. F., Amiot, J., and Bazinet, L. 2006. Simultaneous separation of acid and basic bioactive peptides by electrodialysis with ultrafiltration membrane. *Journal of Biotechnology* 123(3):314–28.

Raghuraman, B., Tirmizi, N., and Wiencek, J. 1994. Emulsion liquid membranes for wastewater treatment: Equilibrium models for some typical metal-extractant systems. *Environmental Science and Technology* 28(6):1090–8.

Ramlow, H., Morais Ferreira, R. K., Marangoni, C., and Machado, R. A. 2019. Ceramic membranes applied to membrane distillation: A comprehensive review. *International Journal of Applied Ceramic Technology* 16(6):2161–72.

Rastogi, N. K. 2016. Opportunities and challenges in applications of forward osmosis in food processing. *Critical Reviews in Food Science and Nutrition* 56(2):266–91.

Romeyn, T. R., Harijanto, W., Sandoval, S., Delagah, S., and Sharbatmaleki, M. 2016. Contaminants of emerging concern in reverse osmosis brine concentrate from indirect/direct water reuse applications. *Water Science and Technology* 73(2):236–50.

Roy, S., and Singha, N. R. 2017. Polymeric nanocomposite membranes for next generation pervaporation process: Strategies, challenges and future prospects. *Membranes* 7(3):53.

Saffarionpour, S., and Ottens, M. 2018. Recent advances in techniques for flavor recovery in liquid food processing. *Food Engineering Reviews* 10(2):81–94.

Saleh, T., and Gupta, V. 2016. *Nanomaterial and Polymer Membranes: Synthesis, Characterization, and Applications*. Amsterdam: Elsevier.

Salgado, C. M., Fernandez-Fernandez, W., Palacio, L., Carmona, F. J., Hernandez, A., and Pradanos, P. 2017. Application of pervaporation and nanofiltration membrane processes for the elaboration of full flavored low alcohol white wines. *Food and Bioproducts Processing* 101:11–21.

Sanaei, P., and Cummings, L. J. 2017. Flow and fouling in membrane filters: Effects of membrane morphology. *Journal of Fluid Mechanics* 818:744–71.

Shon, H. K., Phuntsho, S., Chaudhary, D. S., Vigneswaran, S., and Cho, J. 2013. Nanofiltration for water and wastewater treatment: A mini review. *Drinking Water Engineering and Science* 6:47–53.

Shuit, S. H., Ong, Y. T., Lee, K. T., Subhash, B., and Tan, S. H. 2012. Membrane technology as a promising alternative in biodiesel production: A review. *Biotechnology Advances* 30(6):1364–80.

Strathmann, H. 2011. Membrane separation processes, 1. Principles. In: *Ullmann's Encyclopedia of Industrial Chemistry*. Weinheim: Wiley-VCH.

Strathmann, H. 2001. Membrane separation processes: Current relevance and future opportunities. *AIChE Journal* 47(5):1077–87.

Subramanian, R., Hebbar, H., and Rastogi, N. K. 2007. Processing of honey: A review. *International Journal of Food Properties* 10(1):127–43.

Summe, M. J., Sahoo, S. J., Whitmer, J. K., and Phillip, W. A. 2018. Salt permeation mechanisms in charge-patterned mosaic membranes. *Molecular Systems Design and Engineering* 3(6):959–69.

Synder Filtration 2019. Definition of porous and polymeric membranes. https://synderfiltration.com/learning-center/articles/introduction-to-membranes/polymeric-membranes-porous-non-porous (accessed July 29, 2019).

Terefe, S., Janakievski, F., Glagovskaia, O., De Silva, K., Horne, M., and Stockmann, R. 2016. Forward osmosis: A novel membrane separation technology of relevance to food and related industries. In: *Innovative Food Processing Technologies: Extraction, Separation, Component Modification and Process Intensification*, eds. K. Knoerzer, P. Juliano, G. Smithers, 177–205. Cambridge: Woodhead Publishing.

Tongwen, X., and Weihua, Y. 2004. Tuning the diffusion dialysis performance by surface cross-linking of PPO anion exchange membranes – Simultaneous recovery of sulfuric acid and nickel from electrolysis spent liquor of relatively low acid concentration. *Journal of Hazardous Materials* 109(1–3):157–64.

Turek, M. 2002. Cost effective electrodialytic seawater desalination. *Desalination* 153(1–3):371–6.

Uragami, T. 2017. *Science and Technology of Separation Membranes*. Chichester: John Wiley & Sons Ltd.

Vera, E., Sandeaux, J., Persin, F., Pourcelly, G., Dornier, M., and Ruales, J. 2009. Deacidification of passion fruit juice by electrodialysis with bipolar membrane after different pretreatments. *Journal of Food Engineering* 90(1):67–73.

Wang, L. K., Shammas, N. K., Cheryan, M., Zheng, Y. M., and Zou, S. W. 2011. Treatment of food industry foods and wastes by membrane filtration. In: *Membrane and Desalination Technologies*, eds. L. K. Wang, J. P. Chen, Y. T. Hung, N. K. Shammas, 237–70. New York: Humana Press.

Wang, Z., Ma, J., Tang, C. Y., Kimura, K., Wang, Q., and Han, X. 2014. Membrane cleaning in membrane bioreactors: A review. *Journal of Membrane Science* 468:276–307.

Ward, A. J., Arola, K., Thompson Brewster, E., Mehta, C. M., and Batstone, D. J. 2018. Nutrient recovery from wastewater through pilot scale electrodialysis. *Water Research* 135:57–65.

Warsinger, D. M., Chakraborty, S., Tow, E. W., Plumlee, M. H., Bellona, C., Loutatidou, S., Karimi, L., Mikelonis, A. M., Achilli, A., and Ghassemi, A. 2016. A review of polymeric membranes and processes for potable water reuse. *Progress in Polymer Science* 81:209–37.

Warsinger, D. M., Servi, A., Connors, G. B., Mavukkandy, M., Arafat, H., Gleason, K., and Lienhard, J. H. 2017. Reversing membrane wetting in membrane distillation: Comparing dry out to backwashing with pressurized air. *Environmental Science: Water Research and Technology* 3(5):930–39.

Wee, Y. J., Yun, J., Lee, Y. Y., Zeng, A. P., and Ryu, H. W. 2005. Recovery of lactic acid by repeated batch electrodialysis and lactic acid production using electrodialysis wastewater. *Journal of Bioscience and Bioengineering* 99(2):104–8.

Werber, J. R., Osuji, C. O., and Elimelech, M. 2016. Materials for next-generation desalination and water purification membranes. *Nature Reviews Materials* 1(5):16018.

Williams, C., and Wakeman, R. 2000. Membrane fouling and alternative techniques for its alleviation. *Membrane Technology* 124(124):4–10.

Zhou, J., and Husson, S. M. 2018. Low-energy membrane process for concentration of stick water. *Membranes* 8(2):25.

2

Frequently Used Membrane Processing Techniques for Food Manufacturing Industries

Ulaş Baysan, Necmiye Öznur Çoşkun, Feyza Elmas, and Mehmet Koç

Contents

2.1	Introduction	46
2.2	Advantages and Disadvantages of Membrane Processes	47
2.3	Mechanisms of Membrane Processes	49
2.4	Membrane Processes	51
	2.4.1 Reverse Osmosis	53
	2.4.2 Forward Osmosis	55
	2.4.3 Microfiltration	56
	2.4.4 Ultrafiltration	57
	2.4.5 Nanofiltration	59
	2.4.6 Electrodialysis	60
	2.4.7 Membrane Distillation	61
	2.4.8 Pervaporation	63
	2.4.9 Membrane Bioreactor	65
2.5	Structure, Geometry, and Selectivity of Membranes	66
2.6	Membrane Technology for Food Materials	69
2.7	Application of Membrane Technology in the Food Industry	71
2.8	Conclusion	85
References		86

APPLICATIONS OF MEMBRANE TECHNOLOGY

2.1 INTRODUCTION

Membrane technology has been used in many different fields over the last three decades. Due to the advances in technology, membrane filtration technology has found its application in many industrial sectors such as chemical, petrochemical, mineral, biotechnology, pharmacology, paper, and water treatment (Kuruma and Poetzschke, 2002). The cost of the membrane process is now almost half of the cost paid in the past due to the research conducted by manufacturing companies and the effect of competition. Nowadays, working at low pressures leads to a reduction in energy requirements for the membrane process. Thus, low process costs are one of the factors that expand the use of membrane technology (Hacıfettahoğlu, 2009).

Membrane technology, which does not cause environmental pollution, has gained importance in line with the awareness of environmental issues by society. Membrane science and technology offers new options for the design, rationalization, and optimization of innovative production cycles (Cassano et al., 2011). Thus, it is necessary to first examine what membrane consists of in order to understand, develop, and apply membrane technology, which has many advantages and applications in the food industry.

The word membrane comes from the word *"membrana"* which means "the shell" in Latin (Koçak, 2007). This semi-permeable barrier, which is dependent on selective separation and transport mechanisms, is called the membrane. A membrane is a phase that permits limited and regular passage of one or more species, acting as a barrier against the discontinuity regime or component movement accumulated on the membrane surface between the two phases. In other words, it is an intermediate phase that separates the membrane components according to their structure, size, and chemical structure (Aslan, 2016). The membranes are generally made of metals, organic or inorganic polymers, and they can work as both permeable or semi-permeable (Cheryan, 1998). The use of membranes arose from the need to produce a material with a suitable mechanical strength and a high degree of selectivity, which could provide high separation efficiency (Aslan, 2016). The first function of a membrane is to act as a selective barrier that allows the concentration of one or more components in both the filtrate and the permeate by controlling the transfer of certain components and retention of other components in the mixture (de Morais Coutinho et al., 2009).

Most of the early studies on membrane permeability used natural materials such as animal bladders or rubber for efficient separation

FREQUENTLY USED MEMBRANE PROCESSING TECHNIQUES

(Strathmann et al., 2011). The use of a membrane as a means of separation has been dated back to the 18th century when "Abbe Nollet" first used the word osmosis to describe the water permeability of a diaphragm made from a pig's bladder (Pouliot, 2008). "Abbe Nollet" discovered that when a pig's bladder comes into contact on one side with a water-ethanol mixture and on the other side with water, it was only the ethanol that passed through to the other side (Strathmann et al., 2011). The first plant origin membrane, which was commercialized in the 1930s, was made of nitrocellulose (colodion) with a graded pore size, and it was developed for separation purposes (Pouliot, 2008). In the 1960s, reverse osmosis (RO) was widely used for the small scale desalination of seawater. Forward osmosis (FO) also emerged as a method with a lower energy consumption potential in this decade (Wallace et al., 2008). After the discovery of asymmetric membranes in the early 1960s, membrane technology provided an alternative to clarification and concentration processes commonly used in the dairy and beverage industries (Echavarría et al., 2011). The membrane process has been studied since 1970 to separate microorganisms. Systematic separation techniques are required to collect microorganisms and also to further purify their metabolites. Membrane technology was developed for these processes due to the isothermal process, the non-change of phase, or non-use of chemical substances (Rossignol et al., 1999). When the chronological development of membrane technology is examined, it has been used and developed for different fields since the 18th century. That's why it is important to know the advantages and disadvantages of the membrane process in order to understand where these systems could be applied, and how these systems could be developed and integrated into new process.

2.2 ADVANTAGES AND DISADVANTAGES OF MEMBRANE PROCESSES

Membrane processes, which can be used in many fields and have become increasingly common, have numerous advantages and disadvantages. Because of their lower labor requirements, higher process efficiency, and shorter process time, membrane processes are preferred in many different industries. Also, membrane processes are popular due to their low operating cost (Echavarría et al., 2011). The use of membrane technology offers a wide range of advantages to manufacturers as well as consumers. The advantages of membrane processes in terms of the food industry are low

APPLICATIONS OF MEMBRANE TECHNOLOGY

thermal damage to the product, high aroma retention, low energy consumption, and low equipment costs (Bhattacharjee et al., 2017). Membrane technology is a new non-thermal and environment-friendly technology that minimizes the adverse effects from high temperatures (such as phase change, denaturation of proteins, and changes in sensory properties of the product). In addition, the membranes remove the components that negatively affect the quality of the product, such as microorganisms, drugs, deposits in the product, and also make the final product more attractive as well as increasing the shelf life of the product (Kumar et al., 2013). Separation in membrane systems can be carried out continuously, and easily combined with other separation processes, performed with scaling/grading, and don't require the additives and chemicals (Koçak, 2007). However, the limitations of membrane processes in industrial applications are partly attributable to unfavorable membrane properties such as low permeability, selectivity, thermal, and chemical resistance. An inadequate membrane process design that didn't consider hydrodynamic mechanisms and engineering analysis limited the use of membranes in industry (Drioli and Fontananova, 2004). The general disadvantages of membrane processes are concentration polarization and membrane fouling, low membrane life, and low selectivity or flow (Koçak, 2007).

Concentration polarization and fouling is the main problem with membrane processes. Concentration polarization, which occurs by increasing the concentration of a solution in the membrane boundary region, has been shown to give additional resistance. This resistance increases the operating cost and adversely affects the quality of the filtrate flow. Membrane fouling contributes to pores blockage, clogging, and the formation of cake on the surface of the membrane (Ochando-Pulido et al., 2015). Thus, the fluid flow rate and membrane performance are reduced. These changes also have negative effects on membrane selectivity. The cleaning of membrane systems is also difficult, and the cleaning of in-place membrane systems requires the shut down of the whole process. This being the case, operation and energy costs increase (Cassano et al., 2011; Ochando-Pulido et al., 2015; Turano et al., 2002; Zirehpour et al., 2014). Irreversible fouling leads to a decrease in membrane life. This circumstance reveals an increase in the need for early membrane module replacement and investment cost. Membrane fouling depends on factors such as membrane characteristics, the molecular size of the solution and the interaction of membranes with feed solution properties, and operating conditions (transmembrane pressure, cross-flow rate, operating temperature) (Turano et al., 2002). Research has shown that fouling resistance of the membrane can be improved by smoothing the membrane surface, increasing surface hydrophilicity and providing strong electrostatic repulsion between the membrane surface and charged foulants (Xu et al., 2015; Zhao and Ho, 2014).

Membrane modification is one of the most effective solutions to improve membrane performance. In order to solve the problems occurring in the membranes, the membrane can be modified by adding ingredients with the desired functions, or the surface modification of the membrane necessary to achieve the purpose (Nguyen et al., 2013). The objective of membrane modification is to provide ease of synthesis, reproducibility, high scaling, or high mass potential and the improvement of membrane properties, product quality, permeability, and anti-fouling properties (Goh et al., 2016).

2.3 MECHANISMS OF MEMBRANE PROCESSES

A membrane is a selectively permeable layer that separates two different phases or media (Dizge, 2011). The main mechanism in the membrane is a mass transfer. Mass transfer can be defined as the transfer of one substance from one medium to another. Mass transfer occurs with concentration difference, which transfers substances from the high concentration region to the lower concentration region (Gürel and Büyükgüngör, 2015).

Mass transfer of the membrane is achieved by differences in the physical, chemical, and electrical charge on both sides of the membrane. The differences between the existing energy on both sides results in transport from one side of the membrane to the other side. In other words, a transfer occurs when there is a reduction in the total energy of the system during transport. This difference causes the formation of the energy gradient that is characterized as the driving force. The applied force is in the direction of decreasing energy. There is a decrease in the potential energy of one side of the membrane, where the concentration decreases during separation, due to the mass transfer. This potential energy difference and the thickness of the membrane results in a force, which called the driving force (Aslan, 2016). The general membrane process mechanism is shown in Figure 2.1.

In general, there are two principles for membrane separation technologies. The first principle is the application of a driving force in order to provide flux in the direction of the membrane (Mulder, 2012). Flux can be defined as the amount of material passing through the unit area of the

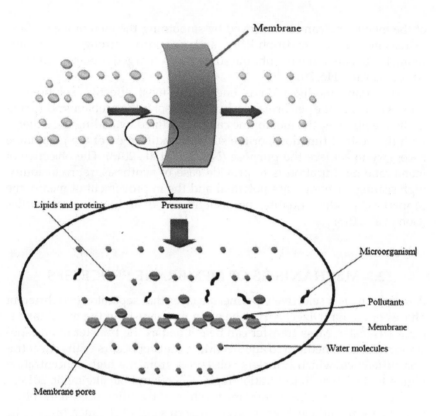

Figure 2.1 Mechanism of the membrane process.

membrane per unit time, which is directly related to the driving force and the total hydrolic resistance of the membrane (Gürel and Büyükgüngör, 2015) The second principle is the separation factor that prevents the passage of certain substances through the membrane (Mulder, 2012).

Membranes allow some of the components in the liquid to pass through depending on their structure and properties, while others do not allow passage and are retained on the membrane surface and/or in the pores. The force must be applied to drive the components through the membranes. The force helps to achieve this separation. The driving force is determined by considering the structure and physicochemical properties of the membranes and the non-transfer components (Aslan, 2016).

The transport of substances from the membrane or mass transfer are carried out utilizing driving forces, such as concentration difference, pressure difference, and electrical potential difference. The pressure difference is one of the most common driving forces applied in membrane processes (Mulder, 2012). Although the kinds of driving forces occur in membrane applications, a pressure difference is the main mechanism in solid membranes, while concentration difference is the dominant driving force in liquid membranes (Gürel and Büyükgüngör, 2015).

Membrane processes can be classified into four groups in terms of the main driving force: reverse osmosis, gas separation, ultra-, micro-, and nanofiltration work based on pressure difference; dialysis and pervaporation work based on concentration difference; electrolysis and electrodialysis work based on the electrical potential difference (Table 2.1). The mechanisms of pressure (ΔP) and concentration difference (Δc) as driving forces are illustrated at Figure 2.2 (a) and (b), respectively, where t is the time in s, P_1 is the feed pressure in Pa, P_2 is the product pressure in Pa, $c_{A,1}$ is the concentration of solute A in feed in kg mol/m^3, $c_{A,2}$ is the concentration of solute A in the product in kg mol/m^3, $c_{w,1}$ is water concentration of feed in kg mol/m^3, $c_{w,2}$ is water concentration of product in kg mol/m^3, N_A is the mass flux of solute A kg mol/m^2.s, N_w is the mass flux of water in kg mol/m^2.s. The mechanism of electric potential difference (ΔE) as driving force is also shown in Figure 2.2 (c), where E_1 is the electrical potential at the cathode side (volt), E_2 is the electrical potential at the anode side (volt), $c_{A,0}$ is the concentration of solute A in feed in kg mol/m^3 at the initial time ($t = 0$), $c_{A,end}$ is the concentration of solute A in the product in kg mol/m^3 at final (t = t_{end}).

2.4 MEMBRANE PROCESSES

There are various membrane processes carried out under the influence of different driving forces. These membrane processes are widely used in different applications for different purposes. Many different membrane

Table 2.1 Classification of Membrane Processes According to Drive Forces

Driving Forces		
Pressure (ΔP)	**Concentration (Δc)**	**Electric Potential (ΔE)**
Microfiltration	Pervaporation	Electrodialysis
Ultrafiltration	Gas separation	Electrolysis
Nanofiltration	Dialysis	
Reverse osmosis	Liquid membranes	

APPLICATIONS OF MEMBRANE TECHNOLOGY

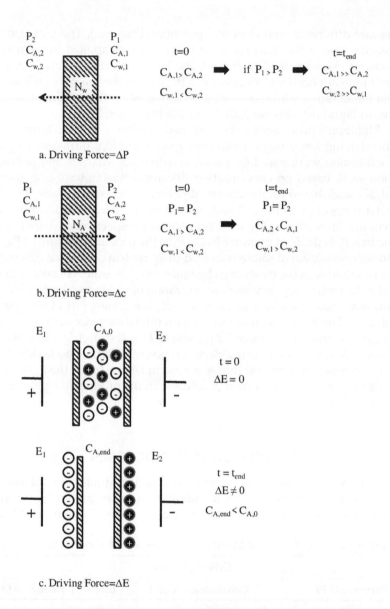

Figure 2.2 Mass transfer in membranes in terms of driving force.

processes exist, and they are generally based on the mass transfer mechanism. The main factors affecting the rate of this mass transfer are the resistance of the membrane, the driving force applied per unit membrane area, hydrodynamic conditions at the membrane–liquid interface, fouling and cleaning of the membrane surface (Judd, 2010).

Reverse osmosis, forward osmosis, microfiltration (MF), ultrafiltration (UF), nanofiltration (NF), electrodialysis (ED), membrane distillation (MD), pervaporation (PV), and membrane bioreactor (MBR) are the main membrane processes used in industry (Koçak, 2007). These methods can be used as a single method or in combination with two or more other methods. In the industrial applications of membrane technologies, the integration of various membrane processes into the same industrial process offers significant benefits in product quality, plant compactness, environmental impact, and energy maintenance (Drioli and Fontananova, 2004).

2.4.1 Reverse Osmosis

Freshwater has to be produced from saltwater and wastewater sources because of an increase in the world population, a higher demand for water resources resulting from industrialization and urbanization, and the limitation of freshwater resources. Therefore, the idea of using reverse osmosis to produce freshwater has emerged and has been widely used (Kang and Cao, 2012; Matin et al., 2011).

RO is a pressure-driven mass transfer mechanism in which a semi-permeable membrane blocks the dissolved components in the feed solution but allows water to pass through the membrane pores (Malaeb and Ayoub, 2011). Although the concept of RO has been known for many years, the usage of RO as a viable separation process is a relatively new technology (Williams, 2003).

Osmotic membrane technologies are dependent on the natural osmosis mechanism, which occurs when two different concentrations of solutions are placed on either side of a semi-permeable membrane. This formation of pressure difference allows the water to move from the dilute solution through the membrane to the concentrated solution, while the selective property of the membrane ensures the retention certian components in the solution (Yip et al., 2010).

In the food industry, the main objective of RO is to concentrate, purify, and recover valuable components. RO helps to reduce or eliminate the operational cost of evaporation (Hedrick, 1983). The formation of undesirable components related to heating are prevented, and the lack of heating costs also provides an additional advantage to RO (Kotsanopoulos and Arvanitoyannis, 2015). Reverse osmosis can also be used in combination with other membrane separation processes such as MF and UF. Other advantages of reverse osmosis are quality separation, minimum heat damage, low waste generation and treatment, smaller footprint, and lower capital requirements (Bhattacharjee et al., 2017).

Reverse osmosis membranes were initially used only for water treatment. However, reverse osmosis is used in many fields, such as bleaching, in various industries (textile, yeast, leather, paper, etc.) in past decades, for industrial wastewater treatment (slaughterhouse, textile, yeast, etc.), the recovery of by-products, leachate treatment, mining, pharmaceutical production, reuse of agricultural drainage water, and recovery of precious metals (Pak, 2011).

RO is used as an alternative technique for juice concentration since it does not require phase change or high temperature. On the other hand, a major disadvantage of RO in the fruit juice industry is that a low concentration of the product is achieved when compared to evaporation by heat treatment. For instance, in conventional fruit juice processing, the concentration levels of the fruit juice range from 42 to 65 °Brix. Therefore, RO can be applied or preferred as a pretreatment with other processes such as osmotic evaporation (Echavarría et al., 2011). However, the major limitation of RO technology in water treatment is membrane clogging, which results in a lower production capacity and higher operating cost (Kang and Cao, 2012).

The most important factors limiting the effectiveness of reverse osmosis are concentration polarization and membrane fouling. A rotating filtration technique can be preferred to overcome these problems. A rotating reverse osmosis system consists of an internal porous cylinder supporting a rotating reverse osmosis membrane in a fixed outer cylindrical shell. The rotating filtration helps to increase slip on the membrane surface. Thus, the effects of concentration polarization and membrane fouling are minimized (Pederson and Lueptow, 2007).

Sugars (hexoses and disaccharides) and organic acids are the main components for providing osmotic pressure in fruit juices. The high pressure (10 to 200 bar) is necessary in order to climb above the osmotic pressure during the juice concentration (Echavarría et al., 2011). Gurak et al. (2010) studied grape juice concentration by reverse osmosis at different pressures (40–60 bar) and temperatures (20–40°C). The high permeate flux rate and 28.5°Brix of juice concentration were obtained at 60 bar and 40°C conditions. Many membrane processes, such as microfiltration, ultrafiltration, and electrodialysis, especially reverse osmosis and nanofiltration, are used in winemaking. Thus, the knowledge in terms of membrane techniques, separation techniques, the membrane, and product characterization should be gathered to provide and develop an effective process. After gathering enough knowledge, the appropriate technique and conditions should be decided carefully (Massot et al., 2008).

An ideal RO membrane is expected to have desirable properties related to high mechanical strength, thermal stability, permeate flux, resistance to dissolution, resistance of the chemical and biological degradation, and long membrane life, as well as a low fouling tendency and cost (Li and Wang, 2010; Misdan et al., 2012; Qasim et al., 2019). However, it may not always be possible to determine whether the membrane has the above-mentioned properties. When this circumstance is faced, it is necessary to determine the best or ideal membrane by considering the purpose, technique, and process conditions to be used.

2.4.2 Forward Osmosis

Clean water, renewable energy, and high medical costs have become major global challenges. Forward osmosis is one of the alternative ways to overcome these challenges. FO can also be used for the dehydration of drugs during synthesis and also for controlled drug release (Chung et al., 2012).

Forward osmosis is a potential alternative method to conventional pressure membrane processes, since hydraulic pressure is not required in this method (Cath et al., 2006). Instead, this process uses a concentrated solution to produce high osmotic pressure, and this pressure is supplied to the water from the surface of the semi-permeable membrane. This solution is then separated from the diluted solution for recycling. Finally, the clean water is obtained (McGinnis and Elimelech, 2007).

FO is based on natural osmosis. Water molecules are transported through a semi-permeable membrane by osmotic pressure difference (Lutchmiah et al., 2014). FO membranes must have high salt retention, high water flux, high resistance to various pH, low concentration polarization, and long-term mechanical and performance stability (Chung et al., 2012).

Forward osmosis, which is also used for seawater desalination processes, works with approximately 20% of the electrical energy required by other desalination processes (microfiltration, ultrafiltration, nanofiltration, reverse osmosis, etc.) (McGinnis and Elimelech, 2007). In addition, the advantage of higher recovery and less saline discharge into the environment is provided by using FO (Cath et al., 2005). The main advantages

APPLICATIONS OF MEMBRANE TECHNOLOGY

of using FO are that it can be carried out at low hydraulic pressure or without hydraulic pressure, it blocks a wide range of pollutants and has lower irreversible fouling than pressure-operated membrane processes due to lack of hydraulic pressure (Lee et al., 2011). Therefore, these advantages lead to a decrease in operating costs. The pharmaceutical synthesis is another area in which advanced osmosis is used. Extraction technologies available for pharmaceutical syntheses consume a large amount of solvent and occupy a large area, and also bring about waste solvent problems. The use of FO technologies for dewatering and concentrating pharmaceuticals can partially solve these problems (Chung et al., 2012; Skilhagen et al., 2008; Thorsen and Holt, 2009).

Despite the advantages of the FO process, there are some problems faced with the use of this technology. These problems are the lower process efficiency, the need for membranes designed specifically for osmotically driven FO processes (commercial semi-permeable membranes are generally designed for pressure-driven processes), and the lack of adequate studies on understanding the contamination and cleaning behavior of FO processes (Mi and Elimelech, 2010).

FO technologies are widely used for seawater desalination, wastewater treatment, and industrial wastewater treatment (Mi and Elimelech, 2008). In addition, forward osmosis technology is utilized to concentrate liquid foods such as fruit juices and milk (Hasanoğlu and Gül, 2016). Hasanoğlu and Gül (2016) studied milk concentration using the forward osmosis method. It has been reported that an increase in the concentration of the draw solution brings about an increase in the difference of water activity on both sides of the membrane. Thus, the increase of the differences in water activity causes an increase in the flux rate.

Moreover, the effect of feed rate was found to be slightly higher than the saltwater rate. This circumstance occurs when the viscosity increases in the feed boundary layer during the process. Hence, it is determined that the resistance of the feed side is more important than the other effects. This study shows remarkably that this technology can be used with the purpose of milk concentration (Hasanoğlu and Gül, 2016).

2.4.3 Microfiltration

The microfiltration (MF) process is the oldest known membrane technology, and its applicability was first demonstrated by Bechold in 1906 (Ripperger and Altmann, 2002). Microfiltration is widely used in potable water treatment as an alternative to conventional water treatment

FREQUENTLY USED MEMBRANE PROCESSING TECHNIQUES

(coagulation, sedimentation, and sand filtration) to remove particles, turbidity, and microorganisms from surface waters and groundwater (Yuan and Zydney, 1999). MF membranes are particularly suitable for the separation of fine particles from 0.1 to 10.0 μm (Kumar et al., 2013). Although the separation mechanism of microfiltration is based on the size difference, the main driving force is also pressure difference in the membrane system. Particles pass directly through the membrane if the diameter of the particles is smaller than the membrane pore diameter whereas larger ones are retained by the membrane. The materials are adsorbed in the pore if the diameter is the same as the membrane pore diameter (Taşıyıcı, 2009). MF applications operate at low-pressure difference (0–2 bar), and the flow direction is applied parallel to the membrane surface. The advantages of the microfiltration system are superior water quality, easier operation control, less maintenance, and low sludge production (Yuan and Zydney, 1999). Microfiltration is the least costly membrane compared to other membranes and usually made of polypropylene, acrylonitrile, nylon, and polytetrafluoroethylene (Ho and Sirkar, 2012).

The concentrated part which cannot pass through the membrane accumulates on the membrane surface and increases the resistance. The biggest problem encountered in MF applications is that the flux decreases with time. The membranes should be cleaned after a certain time to improve flux. Cleaning is carried out with various chemicals. In microfiltration, various methods are applied, such as vortex currents, sound waves, vibrations, electrical fields, etc. to reduce the effect of concentration polarization and membrane fouling (Koyuncu, 2001).

MF membranes are generally used for potable water treatment plants, cooling water, Chlorophyceae removal in lake and fish farms, suspended solids removal from industrial wastewater, and BOD (biochemical oxygen demand) removal from domestic wastewater treatment plants. MF membranes are used for the production of pure water and metal recovery in the semiconductor industry; they are also frequently used in fruit juice, wine, and beer production. The main purpose of MF use in the juice processing industry is to remove suspended solids, fats, and high molecular weight proteins. MF is also used in the dairy industry to purify whey, and degrease and reduce the microbial load of milk (Bhattacharjee et al., 2017; Merin, 1986).

2.4.4 Ultrafiltration

Ultrafiltration (UF) is a membrane separation process that has been used since the 1930s, and it operates at low pressure and is capable of separating

APPLICATIONS OF MEMBRANE TECHNOLOGY

high molecular weight dissolved substances, colloids, microorganisms, and suspended solids from the solution (Pak, 2011). The pore size of the membranes used in UF varies from 0.001 to 0.1 μm. UF membranes consist of two layers where the first layer, with a thickness of 50–250 μm and high permeability, stands on the surface of the membrane and the second layer is supported by this first layer, and it has a high selectivity (Taşıyıcı, 2009). UF membranes are commercially produced in the form of sheets, capillaries, and tubes (Cheremisinoff, 2002). The pressure applied in the UF process is higher than the pressure applied in the microfiltration process and varies in the range of 1–7 bar (Rautenbach, 1997). This pressure is lower than the pressure used in reverse osmosis, but the strength of this pressure cannot overcome the osmotic effects (Koçak, 2007). The UF process can hold molecules with pore sizes ranging from 10 to 1,000 Angstroms and weighing between 300 and 500,000 Daltons (Pak, 2011).

The separation process in membranes depends on the particle size of the material but also on the molecular and colloid forms, the structure of the membrane, and the relationships between the membrane and the entrapped substances. The retention levels of substances are expressed by the molecular weight cut-off (MWCO) and substances below a certain MWCO value cannot be deposited on the membrane (Rautenbach, 1997). The MWCO value defined for each membrane is different. This value can vary depending on the operating conditions, the chemical content of the feed solution, and the molecular properties (Pak, 2011). In addition, the operating temperature of the membranes can be associated with the MWCO. For example, the maximum operating temperature for 5,000 to 10,000 MWCO membranes is 65°C, while the operating temperature for membranes with 50,000 to 80,000 MWCO is around 90°C. Besides the molecular weight inhibition limit, the shape, size, and flexibility of the molecules are other important parameters for the separation process. Molecules that have a very tight structure are removed more successfully than flexible ones for a given molecular weight. Ionic strength and pH are often used for determining the form and firmness of large molecules (Cheremisinoff, 2002; Mulder, 2012; Noble and Stern, 1995).

UF membranes are widely used in the textile industry for recycling sizing agents and indigo dyes; concentrating bleaching wastewater in the paper industry; concentrating oil emulsions in the metal industry; recycling proteins in whey solution in the food industry. Also, UF membranes are used in water samples for the water pollution analysis of concentrated organisms and filtering of cell and cell fractions in aqueous media (Barlas, 2002; Dizge, 2011; Koçak, 2007). In addition, UF is an effective method for

FREQUENTLY USED MEMBRANE PROCESSING TECHNIQUES

separating natural polymers (polysaccharides, proteins) from fruit and vegetable juice (Onsekizoglu et al., 2010). Cassano et al. (2011) recovered the polyphenolic and antioxidant compounds from olive mill wastewaters using UF membranes with different molecular weight and polymeric material. It was observed that cellulose UF membranes contained more polyphenols in the filtrate stream, higher filtrate flow, and lower fouling index than the polyethersulfone membranes (Cassano et al., 2011).

2.4.5 Nanofiltration

Nanofiltration (NF) is an important technology for the food industry, especially for the beverage industry, which requires the reduction of dissolved dirt and fine filtration of the particles (Warczok et al., 2004). NF that has 200–2,000 Da MWCO is a recently introduced membrane separation technique. NF membranes are denser, thinner, and less permeable than other membranes, so operate at higher pressures than MF and UF (10–20 bar) (Akdağlı and Arslan, 2008). Nanofiltration membranes have an asymmetric structure and are a pressure-operated separation process. The membrane pore size varies between 0.5–2 nm and the operating pressure between 5–40 bar. Nanofiltration has different properties, such as pore radius and surface charge density, which affect the separation of solutes (Hussain and Al-Rawajfeh, 2009; Noble and Stern, 1995). The separation process in NF membranes is based on the -hydrodynamic and interface phenomena that occur on the membrane surface and within the membrane pores (Cheng et al., 2011). NF is used to concentrate sugars, divalent salts, bacteria, proteins, and other components with a molecular weight lower than 1 kDa. NF can also be used in part for demineralization (Bhattacharjee et al., 2017). The parameters that affect the performance of NF membranes are pressure, temperature, cross-flow rate, and pH (Hacıfettahoğlu, 2009).

The working principle of NF is similar to reverse osmosis except for the pressure that is applied. The pressure applied is lower than reverse osmosis in the NF process. The energy requirement is therefore lower than reverse osmosis due to the low pressure (Cardew, 2007). The mesh structure of the nanofiltration membranes is clearer than with osmosis membranes. Nanofiltration keeps less single charged ions than double-charged ions. In addition, low molecular weight compounds such as herbicides, insecticides, pesticides, dye, and sugar are also retained at a high rate in the NF membrane. The capital cost of NF membranes is lower than with reverse osmosis since nanofiltration has more water permeability than reverse osmosis (Mulder, 2012). This is why NF is used in the food industry (recycling of nutrients in the fermentation process, purification of organic acids, demineralization of sugar solutions, and separation of amino acids), textile industry (treatment of wastewater from the process), leather industry (recovery of water and salt from wastewater), metal coating industry (recovery of heavy metals from acidic solutions, nickel treatment, etc.) (Ho and Sirkar, 2012).

NF is a process that can especially be used in the fruit juice and sugar industry (Gyura et al., 2002). NF is used for many different purposes, such as the recovery of aroma components, the treatment of wastewater, and the regulation of sugar concentration in the fruit juice industry (Diban et al., 2009; García-Martín et al., 2010; Miyaki et al., 2000). NF can also be used to concentrate the useful bioactive compounds extracted from fruit juices. For example, Arriola et al. (2014) used a nanofiltration membrane to concentrate lycopene from watermelon juice.

2.4.6 Electrodialysis

The concept of electrodialysis (ED) was first proposed in the early 20th century, but the modern ED system was developed in 1950 with the first production of synthetic ion-exchange membranes (IEM) (Juda and McRae, 1950; Teorell, 1935). ED is one of the membrane-based separation processes where ion-exchange membranes are used (Grebenyuk and Grebenyuk, 2002). Electrodialysis is defined as the separation of electrically charged particles from the melt using selected ion membranes. Cations and anions are rejected by the selective membrane. Rejected anions are discharged from the electrodialysis cell (Baker et al., 1990).

The principle of the ED process is based on the migration of ionic species in the aqueous solution, which are exposed to the electric field and contain ions of different mobility, to the corresponding poles of the field. Ionic mobility is directly proportional to the specific electrical conductivity of the solution, while it is inversely proportional to the ionic concentration (Marella et al., 2013).

The anionic and cationic membranes are arranged in a plate and frame configuration like a conventional plate heat exchanger in an ED system. The feed solution is pumped into the cells of the system and an electrical potential is applied. The positively charged ions move toward the cathode, whereas the negatively charged ions move toward the anodes (Marella et al., 2013). Cation exchange membranes (CEM) only transfer cations, while anion exchange membranes (AEM) only transmit anions.

AEM and CEM that placed side by side can remove salt from the solution under an electric field using this ion selectivity (Kwak et al., 2013). Electromotive force is an applied process to provide permeate from the membrane in the electrodialysis (Hasar, 2001).

Electrodialysis has been used on an industrial scale for more than 50 years for the production of drinking water from brackish water sources (Strathmann, 2010). Bacteria, colloidal material, or silica present in the feed water stream remain in the product stream and cause contamination because the ED only removes ions. The polarity of the system can be reversed by reversing the electrodialysis (reverse-ED) in order to minimize contamination and thus the need for the addition of chemical products. The ions are moved in the opposite direction through the membranes, and accumulation is minimized by reversing the polarity (and the flow direction of the solution) several times per hour. Reverse-ED does not directly filter flow through the membranes; contaminants are discharged from the process stream and held by the membranes (Bernardes et al., 2016).

Recently, electrodialysis combined with bipolar membranes or ion-exchange resins has been applied in the chemical, food, and pharmaceutical industries, as well as in wastewater treatment and in the production of high-quality industrial water (Strathmann, 2010). It is also widely used for the demineralization of liquid foods such as milk and whey and for the desalination of seawater (Marella et al., 2013), deacidification of fruit juices, pH control, and heavy metal recovery, and the production of caustic soda in chlor-alkali plants (Salt et al., 2006).

The basic control parameters of an electrodialysis process are electrical conductivity, pH, and electrolyte concentration, electric current, or applied potential. However, it is necessary to understand the electrochemical behavior, stability, conductivity, and selectivity of ion-exchange membranes in order to ensure the efficient use of technical feasibility and electrodialysis processes. It is also important to observe certain specific electrochemical processes, such as the occurrence of concentration polarization and limiting current density (Bernardes et al., 2016).

2.4.7 Membrane Distillation

Membrane distillation is a membrane technology based on the vapor pressure gradient across the porous hydrophobic membrane. Only volatile molecules can be transported through the membranes, and the feed fluid directly in contact with the membrane is not allowed to pass through

APPLICATIONS OF MEMBRANE TECHNOLOGY

the pores of the membrane (Wang and Chung, 2015). The hydrophobic nature of the membrane prevents mass transfer in the liquid phase and forms a vapor–liquid interface at the inlet of the pore. Volatile compounds that present through the membrane pores are condensed and/or removed from the opposite side of the system (Curcio and Drioli, 2005).

The driving force for MD processes is the vapor pressure difference across the membrane rather than an applied absolute pressure difference, a concentration gradient, or an electric potential gradient that drives mass transfer from a membrane (Camacho et al., 2013). Four basic MD configurations are available. The first and most common of these configurations is direct contact membrane distillation (DCMD). When the hot and cold aqueous solution is separated by a non-wetting membrane, the water vapor diffuses from the hot solution interface to the cold solution interface and condenses there. The pressures on both sides may be different unless the membrane pores are wetted with both solutions. In this case, the microporous membrane acts as a liquid phase barrier, while water evaporation continues. This is called direct contact membrane distillation (Farid, 2010). In direct contact membrane distillation, the penetration side of the membrane consists of a condensing fluid (usually pure water), which is in direct contact with the membrane. Both feed and permeate streams can be operated in countercurrent flow, and relatively high heat transfer coefficients can be achieved and horizontal or vertical mounting of the membrane modules can be realized. Other alternative configurations are air gap membrane distillation (AGMD), in which the vaporized solvent can be separated from the membrane by an air gap, vacuum membrane distillation (VMD) in which it can be separated by vacuum (VMD), and sweep gas membrane distillation (SGMD), which can be separated by a sweep gas and recovered on a condensate surface. The air gap configuration (AGMD) can be widely used, especially for many membrane distillation applications where energy availability is low. Vacuum membrane distillation (VMD) and sweep gas membrane distillation are preferred when volatiles are removed from an aqueous solution (Camacho et al., 2013).

The main advantages of membrane distillation are the possibility of entrainment, the possibility of horizontal configuration, the use of low-temperature energy sources, the reduction of contamination problems by the use of hydrophobic membranes, and the possibility of a very compact design, such as a hollow fiber configuration (Farid, 2010). MD provides 100% (theoretical) rejection of inorganic ions, macromolecules, and other non-volatile compounds, uses relatively low operating temperatures and operates at lower operating pressures than conventional pressure-operated membrane separation processes. Also, it is insensitive to feed concentration in seawater desalination, and there is less need for membrane mechanical properties (Wang and Chung, 2015). Membrane distillation was discovered in the late 1960s. However, it was not commercialized for desalination processes due to the lack of membranes and high processing costs compared to reverse osmosis (Camacho et al., 2013). Membrane distillation was used for the removal of heavy metals from wastewater (Zolotarev et al., 1994), recovery of HCl from the cleaning solution in electrolysis (Tomaszewska et al., 2001), concentration of sulfuric acid for the recovery of lanthanide compounds in the extraction process of apatite phosphogypsum (Tomaszewska, 1993), elimination of radioisotopes, reduction of the volume of waste from the nuclear industry (Zakrzewska-Trznadel et al., 1999), and removal of volatile organic compounds from dilute aqueous solutions (Banat and Simandl, 1996; Semmens et al., 1989).

MD can be used in the fruit juice and dairy industries. This method is superior to thermal methods because it provides the concentration of fruit juices at low temperatures. The main advantage of concentrating fruit juice at low temperatures is that the aroma and organoleptic properties of the product are preserved (Bandini and Sarti, 2002; Varming et al., 2004). In addition, evaporation of whey and skimmed milk can be performed at milder temperatures by MD (Christensen et al., 2006; Hausmann et al., 2011).

Laganà et al. (2000) produced apple juice using direct contact membrane distillation and a hollow fiber module. It was found that the transmembrane motion force decreased with increasing membrane temperature but increased with higher feed and distillate flow rates. They also reported that the viscosity of the juice at high concentrations caused a high polarization (Laganà et al., 2000).

2.4.8 Pervaporation

Pervaporation (PV), where separation is performed with a polymeric membrane, is an effective process for separating or recovering components from organic–water or organic–organic mixtures which are difficult to separate or require high energy costs in conventional separation processes (Salt et al., 2006). The history of the pervaporation process dates back to the 1910s when "Kober" defined the term "per-vaporation." The term pervaporation was created by shortening the words "permeation" and "evaporation," after observing the selective permeability of water from the colodium and parchment membrane (Kober, 1917). The liquid mixture (feed) to be separated is placed in contact with one side of the membrane, and the passed mixture (penetration) is removed as a low-pressure vapor on the other side in the pervaporation process (Feng and Huang, 1997). Partial evaporation through the membrane is responsible for the separation potential of the pervaporation. The process results in a vapor permeate and a liquid retentate (Karlsson et al., 1995). Permeate stream may be condensed and collected or released as desired (Feng and Huang, 1997). The most important factor that distinguishes pervaporation from other membrane processes is the phase change during the separation process (Jiraratananon et al., 2002; Pacheco and Marshall, 1997). The chemical potential gradient across the membrane is the driving force for mass transfer (Feng and Huang, 1997). The penetration of a component through the pervaporation membrane can be explained both thermodynamically and kinetically. Thermodynamically, it contains the solubility of the component in the membrane material, while kinetically, it allows the diffusion of the penetrant across the membrane (Huang, 1991).

Pervaporation has many different operating modes. Vacuum pervaporation, also called standard pervaporation, is the most widely used mode of operation. Inert purge pervaporation can be used if the permeate can be drained without condensation. In addition, thermal pervaporation, injury or osmotic distillation, saturated vapor permeability, and pressure-driven pervaporation are some of the modes of operation for pervaporation (Franken et al., 1990; Gonçalves et al., 1990).

PV is a diffusion-controlled process because the flux is generally low. Therefore, the process is more economical when the components that selectively pass through the membrane have a low feed flow rate (Shah et al., 1999). This can be solved by reducing the effective membrane thickness and increasing the membrane area packing density (Feng and Huang, 1997). PV applications are also referred to as clean technology. PV process can be divided into three categories: removal of water from organic solvents, removal of organic components from aqueous solutions, and separation of organic mixtures (Smitha et al., 2004).

Pervaporation is one of the most effective membrane processes for aroma recovery in beverages and has been used in recent years to recover aroma components from fruit juices. Pervaporation membranes are very selective of several components that are important in the aroma profiles of beverages (Olmo et al., 2014). In addition, pervaporation is used for ethanol removal from alcoholic beverages (Takács et al., 2007; Verhoef et al., 2008) and aroma gain (Brazinha and Crespo, 2009; Karlsson et al., 1995), wine dealkalization (Catarino and Mendes, 2011), and non-alcoholic beer development (Gudernatsch and Kimmerle, 1991).

Catarino et al. (2009) extracted and analyzed seven aromatic compounds (ethanol, propanol, isobutanol, isoamyl alcohol, ethyl acetate, isoamyl acetate, and acetaldehyde) in the aroma profile of the beer by pervaporation. The alcohol/ester ratio was reported to be directly proportional to temperature while inversely proportional to the feed rate and pressure (Catarino et al., 2009).

Pervaporation can be used as an alternative to conventional separation processes such as steam distillation, liquid solvent extraction, and vacuum distillation. The advantage of this method is that energy consumption is generally lower, no chemical additives are needed, and it can be operated at the low temperatures when separating sensitive aroma compounds (Olmo et al., 2014).

2.4.9 Membrane Bioreactor

Membrane bioreactor (MBR) technology is a combination of micro and ultrafiltration with an activated sludge process. A membrane bioreactor is an effective method of choice for industrial water treatment and water reuse due to its high product water quality and low footprint (Hoinkis et al., 2012; Judd, 2010).

The pore diameter of the membranes is between 0.01–0.1 µm. Particles and bacteria are retained on the pore, and the membrane system is used instead of the conventional gravity sedimentation unit (scrubber) in the biological sludge process. Thus, the method offers higher water quality and lower footprint advantage (Hoinkis et al., 2012; Lesjean and Huisjes, 2008).

MBR systems are used in two main configurations: external (where liquid and biomass separation occurs in a separate unit by cross-flow membrane filtration) and internal-integrated (where liquid and biomass separation occurs with submerged membranes in the bioreactor). Submerged MBRs require less energy. However, submerged MBR systems are operated with lower permeation fluids since they provide low membrane surface cutting levels. This being the case more membrane surface is required. Today, most commercial applications rely on submerged configurations due to their low energy requirements. Also, submerged MBR systems are preferred because of their robustness and flexibility (Aslan, 2016; Judd, 2010). Advantages of membrane bioreactor systems are their high efficiency in organic material removal, stable nutrient separation, their ability for difficult waste treatments, low sludge production, high-quality disinfection of effluent, high loading speed, less contaminated sludge formation, and a small area for the reactor (Çinar et al., 2006).

The main constraints to the implementation of MBR systems are related to the initial investment costs of the plants and the variable costs, such as electricity consumption and the running time of the membranes. This is a disadvantage for the application of this technology, especially in the treatment of large amounts of wastewater. On the other hand, the use of MBR is more appropriate in the case of stricter discharge rules or when water needs to be treated (Artiga et al., 2005). Another major limitation associated with MBRs is membrane obstruction. Fouling reduces permeable flux and increases the frequency of membrane cleaning and replacement (Le-Clech et al., 2006). Clogging reduces production and increases operating and maintenance costs; in particular, the need for extra cleaning and backwashing, transmembrane pressure (TMP) increases to maintain constant flux. Obstructions may occur on the membrane surface or within the membrane pores. In addition to wastewater treatment, membrane bioreactors can be used in a wide variety of fields, such as the production of foods, plant metabolites, amino acids, antibiotics, anti-inflammatories, anticancer drugs, vitamins, proteins, optically pure enantiomers, isomers, fine chemicals, and biofuels (Cassano et al., 2011).

2.5 STRUCTURE, GEOMETRY, AND SELECTIVITY OF MEMBRANES

The structure, geometry, and selectivity of the membranes are the main factors determining the usage purpose and fields. Not all membrane types can be used for every membrane technique, so the appropriate membrane should be examined and selected in detail. The selectivity of the membranes also plays a critical role in determining the appropriate membrane. Membranes serve as barriers to the selective separation and transport of specific components. The separation process is determined by both the chemical and physical properties of the membrane, and this process is carried out with the help of the driving force (Singh, 1998). The selectivity of the membrane relates to the separated molecule or particle, such as separation and pore size, as well as the diffusivity of the solute (de Morais Coutinho et al., 2009). The separation process is controlled by size, shape, and charge separation for porous membranes, while the separation process using non-porous membranes depends on a sorption and diffusion model (Singh, 1998). Selectivity is impaired by large diameter pore size distribution. The optimal physical structure for any membrane material depends on the small pore size and the thickness of the material that has a high level of porosity. Membrane permeability increases with increasing pore density.

Moreover, membrane resistance is directly proportional to membrane thickness (Aslan, 2016). However, the separation performance of a membrane is controlled by the membranes' chemical composition, operating temperature and pressure, and interactions between the components in the feed solution and the membrane surface properties (de Morais Coutinho et al., 2009). In other words, the performance of membrane processes depends on the structure and morphology of the membrane. Therefore, the performance of a membrane varies considerably depending on the type of membrane used (Dizge, 2011).

A wide variety of membranes are used in membrane applications, and these membranes consist of a solid, dense, or porous polymer, ceramic or metal films with symmetrical or asymmetric structures, and liquid films with selective carrier components and electrically charged barriers (Strathmann, 2001). A membrane can be thin or thick, homogeneous or heterogeneous; it can perform active or passive transport. Active transport can perform with the main driving force outside, whereas passive transport can be realized by pressure, temperature, or concentration difference. In addition, the membrane may be synthetic, natural, porous or non-porous, organic or inorganic, neutral or charged (Koçak, 2007). The choice of membrane material depends on the purpose and requirement of the process, but these properties may not always be sufficient to achieve the best performance (Goh et al., 2016). Thus, it is important to evaluate all mentioned membrane properties together.

Membranes can be classified in many ways according to their separation mechanisms, morphology, geometry, and chemical structure properties. Membranes are classified as organic and inorganic membranes according to their chemical structure. Organic membranes are produced from polymers, but polymer membranes are not preferred for membrane processes because of the short lifetime and process requirements of the membrane. Commercially, thin-film composite (TFC) polyamide (PA) and cellulose acetate (CA) membranes are widely used for RO and FO, while hydrophobic polytetrafluoroethylene (PTFE) and polyvinylidene-fluoride (PVDF) are widely used in MD (Goh et al., 2016).

Inorganic membranes are more resistant to chemical, thermal, and mechanical processes than organic membranes, so inorganic membranes can be applied at high pressure and works in a wide range of pH (Koyuncu, 2001). However, the commercial usage of inorganic membranes is limited because of lower durability and higher cost (Zhou and Smith, 2002). Inorganic membranes can be classified as ceramic, glass, and metallic membranes according to their main construction materials (Koyuncu, 2001). Furthermore, membranes divide into symmetrical, asymmetrical, and thin-film composite membranes according to their morphology. Symmetrical membranes are homogenous due to being produced by a single material, and the thickness of symmetrical membranes (with and without cavities) varies between 10–200 µm. The pores of the membrane have an equal size in each region, and almost all pores have a fixed diameter along the lateral section of the membrane (Koyuncu, 2018). Asymmetric membranes have a non-uniform structure and can be both homogeneous and heterogeneous. The thickness of the asymmetric membranes that contain a dense top layer of 0.1–0.5 µm thickness and a substrate of 50–100 µm thickness varies between 10–200 µm (Mulder, 2012). The active surface layer of asymmetric membranes is generally composed of cellulose acetate or polyamide, and the support layer consists of polysulfone or polypropylene (Crespo and Böddeker, 2013). Asymmetric membranes provide not only better separation performance than symmetric membranes, but also better permeability levels (Koyuncu, 2018).

Thin-film composite membranes have two different layers, which are obtained by placing a thin layer on the top of the asymmetric membranes. This is why composite membranes have a heterogeneous structure. One of the layers is porous and forms the support layer; the other is a nonporous layer and forms the top layer (Koyuncu, 2018). The thickness of the top layer is 1% of the total membrane thickness, so this layer called a thin layer. These types of membranes are used in microfiltration, ultrafiltration, nanofiltration, and reverse osmosis systems. The microfiltration and ultrafiltration membranes are symmetrical or asymmetric, while nanofiltration and reverse osmosis membranes are asymmetrical (Mulder, 2012).

Solid and liquid membranes are used in water and wastewater treatment. The separation process duration is different in solid and liquid membranes. The phase to be purified from one side of the membrane enters the system in solid membranes. A filtrate stream that can pass through a solid membrane and concentrated stream, which cannot pass through the solid membrane, leaves the system. The treated water cannot pass through the liquid membrane, but the chosen substances can pass through (Giwa and Hasan, 2015).

In all the above-mentioned membrane types, the direction of application through the surface of the membrane varies. There are two filtration methods applied in the operation of membrane processes: normal-flow filtration (dead-end filtration) and cross-flow (tangential flow) filtration

(Aslan, 2016). In normal-flow filtration, the fluid is fed directly to the entire surface of the membrane from one side, and substances present in the water, which are larger than the pore size of the membrane, form a layer of cake on the membrane surface (Tansel et al., 2006). The whole solution passes through the membrane, and only one flux is generated (Aslan, 2016). In normal-flow filtration, there is no retentate flow. In order to prevent the formation of this thick cake layer and to maintain high flux velocities, turbulence is formed on the membrane by mixing the fluid frequently (Tansel et al., 2006).

In the cross-flow mode, the fluid to be filtered flows parallel (tangential) to the membrane surface (Vigneswaran et al., 2005), and it is filtered through the membrane due to the pressure difference. The main advantage of this operating mode is that it reduces filter cake formation. The filter cake formation occurs at the cross-flow, which is at a lower level than the normal flow (Charcosset, 2006). Since there is a continuous flow on the membrane surface, the concentrated part (components that do not pass through the membrane) is continuously removed from the surface (Aslan, 2016). The problem of contamination in both cross-flow and normal flows causes significant limitations. However, the superficial flow in the cross-flow mode allows the reduction of contamination and allows for more favorable flux values (Cassano and Drioli, 2013).

2.6 MEMBRANE TECHNOLOGY FOR FOOD MATERIALS

Membrane technology is a separation process that can be used in different industries such as the chemistry, pharmaceutical, biotechnology and food industries (de Morais Coutinho et al., 2009). Membrane processes are preferred for the food industry because of reduced manpower requirements, it is more efficient, it can be carried out at low temperatures, it gives less damage to the product, and it has a shorter processing time than conventional processes (Bhattacharjee et al., 2017; Yazdanshenas et al., 2010). Membrane technology is widely applied in the dairy and beverage industry, especially for fruit juice clarification and concentration (Nunes and Peinemann, 2001).

Raw fruit juice contains significant amounts of macromolecules such as polysaccharides (pectins, cellulose, hemicelluloses, and starch), turbidity particles (suspended solids, colloidal particles, proteins, and polyphenols) and low molecular weight components such as sugar, acid, salt, flavor, and aroma compounds. Therefore, fruit juice should be clarified for a longer storage capability. Clarification is usually carried out using

APPLICATIONS OF MEMBRANE TECHNOLOGY

membranes as well as fining agents (gelatin, bentonite, etc.) (Rodrigues and Fernandes, 2012).

Integrated membrane processes have significant potential for the production of high-quality concentrated fruit juices (Drioli and Fontananova, 2004). Clogging of membranes is the major problem in the use of membrane systems for clarification and/or concentration of fruit juices. Clogging is caused by a decrease in flux due to the operating time and reduces membrane permeability. The degree of fouling of the membrane determines the frequency of cleaning, the life of the membrane, the required membrane area, and consequently the design and operation of the membrane plant (Bhattacharjee et al., 2017). Fruit juices have a high solid and pectin content, so when direct reverse osmosis or osmotic distillation is applied the permeate flux is very low.

Furthermore, fruit juice cannot be concentrated greater than 25–30°Brix due to the high osmotic pressure limitation of a single-stage reverse osmosis system. The use of microfiltration or ultrafiltration as a pretreatment reduces the viscosity of the flow and increases the flux. This ensures that suspended solids and pectin are separated from the water. The volatile components are retained in fruit juice at a higher rate, and a better quality product is obtained by integrated membrane processes (Drioli and Fontananova, 2004).

The milk and dairy industry is another application area for membrane technology in the food industry. Milk is generally concentrated using multi-stage evaporators to ensure microbiological and chemical stability. This concentration process is also an important step in the production of various dairy products such as cheese, yogurt, and milk powder (Van Den Berg, 1962). Ramírez et al. (2006) performed an analysis of energy consumption and energy efficiency for the dairy industry. It was reported that milk concentration is the most intensive energy-consuming operation of the dairy industry. Therefore, the use of membrane separation techniques in a liquid nutrient concentration has many advantages over conventional separation processes, including improved product quality, easy scaling, and low energy consumption (Jiao et al., 2004). Membrane separation technology can also be used for the recovery of whey in the dairy industry. Many whey components can be separated on a size basis. The separation process is carried out without phase transition. Recycling of small quantities and valuable components from a mainstream using the membrane can be performed without significant additional energy costs (El-Sayed and Chase, 2011; Takht Ravanchi et al., 2009). It is estimated that most ultrafiltration membranes in the dairy industry are used to separate whey proteins (Saltık et al., 2017).

The use of membrane technology is seen as an important alternative to traditional methods of processing vegetable oils. The wastewater that comes from the olive oil process is nowadays treated with membrane technology. This wastewater contains organic materials, phytotoxic substances, and various phenol and phenolic compounds that are not readily biodegradable. Discharge of olive oil wastes to the environment causes negative environmental effects such as aquatic life-threatening, odor problems, and toxicity. El-Abbassi et al. (2014) used a polyethersulfone membrane that had a molecular weight of 50 kDa for the treatment of olive processing wastewater and found that the UF membrane was an effective pretreatment. Garcia-Castello et al. (2010) indicated that 91% of suspended solids and 26% of total organic carbon were removed from the olive mill wastewater by MF pretreatment. The different membrane processes in food applications are summarized in Table 2.2 and Table 2.3.

The potential of membrane processes in different areas of food production is well known. The new processes depend on the combination of different membrane processes or the combination of membrane processes and conventional separation technologies (i.e., adsorption, centrifugation, evaporation). These combinations offer interesting benefits when developed as a single concept and offer interesting insights into the revision of traditional technologies in food processing. In addition, the development of hybrid processes offers new and more opportunities for product quality improvement, process or product innovation (Cassano et al., 2001). Some different studies on hybrid membrane processes used in the food industry are given in Table 2.3.

2.7 APPLICATION OF MEMBRANE TECHNOLOGY IN THE FOOD INDUSTRY

The application of membrane technology in the food industry has been increasing in the market since the 1960s. Membrane technology has many advantages for the food industry, such as high separation accuracy, better selectivity, the ability to operate at a lower temperature, high automation, easy operation, lower operating cost, and lower chemical degradation. Thus, many companies in the food industry use membrane processes on an industrial scale and they also provide financial support to pilot and laboratory-scale academic studies in this field. The main membrane applications are the removal of microorganisms from food; purification or concentration of alcoholic and non-alcoholic beverages; concentration or separation-purification of proteins; purification of oil and wastewater

Table 2.2 Earlier Studies on Membrane Technology in Food Process

Aim of Process	Membrane Process	Material	Process Condition	Key-results	References
Clarification	Ultrafiltration	Apple juice	50°C	• Color of apple juice was improved and some minor components, especially phenolics and trans-2-hexenal were retained by ultrafiltration.	Onsekizoglu et al., 2010
Clarification	Ultrafiltration	Apple juice	150–400 kPa 2–7 m/s	• Physical properties like color, clarity and turbidity were improved by ultrafiltration. • Color, soluble solids, pH, and acidity of juice kept stable for 6 months at 16°C.	de Bruijn et al., 2003
Concentration	Reverse Osmosis	Apple juice	5.50 MPa 25°C	• It resulted that the solution-diffusion model combined with the film model was performed successfully to fit the reverse osmosis process in apple juice concentration.	Alvarez et al., 1997
Clarification	Ultrafiltration	Kiwifruit juice	0.85 bar 800 L/h 25°C	• The UF process provided a good clarification process by reducing the suspended solids and the turbidity of the fresh juice.	Cassano et al., 2007

(*Continued*)

Table 2.2 (Continued)   Earlier Studies on Membrane Technology in Food Process

Aim of Process	Membrane Process	Material	Process Condition	Key-results	References
Concentration	Reverse Osmosis	Grape juice	20, 30, and 40°C 40, 50, and 60 bar	• It was concluded that the reverse osmosis process could be considered as a pre-concentration process, which could be combined with other technologies to concentrate fruit juices up to 60 Brix as required and prevents quality loss.	Gurak et al., 2010
Membrane orientation	Reverse Osmosis	Grape juice Beetroot juice Pineapple juice	6.0 M or 26% w/w	• It was observed that further reverse osmosis was to be potentially useful for the concentration of fruit juices, especially if the juice contains various sugars and heat-sensitive compounds such as betalain or anthocyanin.	Nayak et al., 2011
Filtration	Reverse Osmosis	Orange juice	25°C	• The low temperature and pressure applied in the process was an important criterion for the preservation of the sensory properties of the beverage. • Feed stream was refrigerated in each concentration stage, so in high retention of the flavor and color of the juice concentrate.	Beaudry and Lampi, 1990

(Continued)

Table 2.2 (Continued)   Earlier Studies on Membrane Technology in Food Process

Aim of Process	Membrane Process	Material	Process Condition	Key-results	References
Concentration	Reverse Osmosis	Pineapple juice	100 Da 25–45°C	• It was concluded that the use of combined solutes could result in better sensory quality of the product for the direct osmosis process. • Using mixed osmotic agent was found to be most preferable for concentrated pineapple juice according to the sensory analysis results.	Babu et al., 2006
Cold sterilization Clarification	Microfiltration	Pineapple juice	25°C 100 kPa	• Pineapple juice was successfully clarified by microfiltration. • The clarified pineapple juice also satisfied the "commercial" sterilization grade, which described as the requirements of microbiological safety.	Carneiro et al., 2002
Clarification and stability	Nanofiltration	Pear juice	12.4 bar, 50°C	• Applying nanofiltration resulted in lower sugar content, lighter color, and lower organic acids content in the final product.	Vivekanand et al., 2012

(Continued)

Aim of Process	Membrane Process	Material	Process Condition	Key-results	References
Concentration	Reverse Osmosis	Blackcurrant juice	25°C 60 bar	• It was found that reverse osmosis was a suitable method for the concentration of blackcurrant juice up to 28.68 Brix.	Pap et al., 2009
Clarification	Microfiltration	Sherry wines	4 to 50°C 11×104 Pa	• The results showed that the microfiltration has higher effectiveness than conventional filtration for the clarification of the sherry wines.	Palacios et al., 2002
Stabilization	Electrodialysis	Sherry wines		• It was found that, if the value of the electrodialysis of 30% was not exceeded, Reductions of the aroma and flavor were acceptable.	Gómez Benítez et al., 2003
Dealcoholization	Reverse Osmosis	Homemade alcoholic beverages	35–50 bar 0°C	• The reverse osmosis was not economically suitable for the production of beverages has an alcohol percentage lower than 0.45%.	Pilipovik and Riverol, 2005

(Continued)

APPLICATIONS OF MEMBRANE TECHNOLOGY

Table 2.2 (Continued) Earlier Studies on Membrane Technology in Food Process

Aim of Process	Membrane Process	Material	Process Condition	Key-results	References
Stabilization	Electrodialysis	Wine tartaric	25°C 0.2 g/1 NaCl solution	• The electrodialysis that used to extract the ions allows removing the potassium hydrogen tartrate to any required extent to obtain a wine of the desired tartaric stability.	Gonçalves et al., 2003
Separation	Ultrafiltration	Apple pectin	0.02–0.1 MPa 2–7 m/s 70°C	• Pectin isolation by ultrafiltration had a better effect for discoloration of pectin solution because of the absorptive function of ultrafiltration for pigments and impurities.	Qiu et al., 2009
Concentration	Reverse Osmosis	Blackcurrant	30°C 30–50 bar	• Reverse osmosis, which used to filtration at low temperature, preserved the valuable components (anthocyanins, phenols, acids, etc.) and the antioxidant capacity of the blackcurrant during the process.	Bánvölgyi et al., 2009

(Continued)

FREQUENTLY USED MEMBRANE PROCESSING TECHNIQUES

Table 2.2 (Continued) Earlier Studies on Membrane Technology in Food Process

Aim of Process	Membrane Process	Material	Process Condition	Key-results	References
Concentration	Reverse Osmosis	Garcinia indica Choisy (kokum)	1.0–6.0 MPa 25°C	• These results clearly indicated that forward osmosis that used to concentration of anthocyanin extract had several advantages in terms of lower browning, higher stability index according to thermal concentration.	Nayak and Rastogi, 2010
Reduction of microbial content	Microfiltration	Bovine milk Ovine milk	≤ 40°C pH: 6.41–6.5	• The filtration step before microfiltration was demonstrated to be necessary for high performance.	Beolchini et al., 2004
Concentration	Nanofiltration	Coffee extract	28 and 40 bar 40°C	• It was concluded that nanofiltration membranes were capable of concentrating coffee extract to a certain level. • Nanofiltration membranes have the potential to partially replace the conventional evaporation process.	Pan et al., 2013
Concentration	Ultrafiltration	Yogurt	30 kDa 50°C	• It was concluded that performing ultrafiltration on milk before fermentation for the manufacture of yogurt was beneficial in terms of mass balance.	Paredes Valencia et al., 2018

(Continued)

Table 2.2 (Continued) Earlier Studies on Membrane Technology in Food Process

Aim of Process	Membrane Process	Material	Process Condition	Key-results	References
Separation	Electrodialysis	Aqueous solutions	33°C pH: 3–7	• Electrodialysis appears to be highly effective in separating sodium citrate from aqueous solutions, but the operating costs of this technique were found to be about 50% higher than the costs of the current industrial-scale citric acid recovery process.	Moresi and Sappino, 1998
Concentration	Nanofiltration	Grape must	41 bar 15°C	• Nanofiltration helped to increase the sugar content of grape must be used in wine processing with limited alternation in malic and tartaric acid content.	Versari et al., 2003

Table 2.3 Earlier Studies on Hybrid Membrane Process Used in Food Materials

Aim of the Process	Hybrid Membrane Process	Material	Process Condition	Key-results	References
Clarification, Concentration	Membrane Bioreactor Reverse Osmosis Pervaporation	Apple Juice	**MB:** 50°C for 2 h or at 20°C for 8–12 h **RO:** 70 bar, 25°C, 4,000 l/h **PV:** 2 mbar, liquid nitrogen at –196°C or with dry ice at –56°C.	• Pervaporation was found to be suitable for aroma recovery from apple juice. • The organoleptic evaluation of cleaned and concentrated apple juices was excellent in odor and flavor. • Apple juice was cleaner and lighter than those produced by traditional methods. • The integrated membrane process showed comparable than the traditional method.	Alvarez et al., 1997
Concentrated Clarified	Reverse Osmosis Microfiltration Osmotic evaporation	Apple juice	**RO:** 25°C, 6 MPa **OE:** 30°C, 20 kPa **MC:** 2 bar, 30°C	• The results of the study showed that osmotic evaporation could be a promising method for obtaining high-quality low volume concentrated products. • Sensory evaluation showed that the reconstituted concentrated juices were accepted by consumers, although the juice lost some volatile compounds.	Aguiar et al., 2012

(*Continued*)

Aim of the Process	Hybrid Membrane Process	Material	Process Condition	Key-results	References
Concentration	Osmotic evaporation (OE) Membrane distillation	Orange juice	**OE:** 4.9 M, 25°C **MD:** 34°C, 23°C	• It was concluded that the osmotic evaporation process had advantages in comparison to the membrane distillation process, not only in terms of water flux stream, but also regarding the retention of aroma compounds. • As regards the transport of aroma compounds through the membrane, higher retention was observed for each dewatering amount in the osmotic evaporation process.	Alves and Coelhoso, 2006
Concentration	Ultrafiltration Nanofiltration	Bergamot juice	**UF:** 0.7 bar, 114 1/h, 24°C **NF:** 0–60 bar	• An integrated UF-NF system was a suitable approach for the recovery of bergamot juice in the form of phenolic fraction for use as a functional component.	Conidi et al., 2011

(Continued)

Table 2.3 (Continued) Earlier Studies on Hybrid Membrane Process Used in Food Materials

Aim of the Process	Hybrid Membrane Process	Material	Process Condition	Key-results	References
Concentration	Membrane distillation Reverse osmosis	Grape juice	**MD:** 15°C, 30°C **RO:** 60 w/w $CaCl_2$ ~140 bar	• High sugar concentration in grape juice could be reached at low temperatures by using microfiltration and reverse osmosis.	Kozák et al., 2006
Clarification Concentration	Ultrafiltration Osmotic Distillation Pervaporation	Kiwifruit juice	**UF:** 15 kDa, 40°C **OD:** 28 l/h, 0.28 bar **PV:** 5 mbar, 3 l	• It was concluded that the incorporation of pervaporation into the integrated process may prevent the loss of aroma compounds from the kiwi juice prior to ultrafiltration and osmotic distillation. • The results showed that the recovered aroma components could be added to the final concentrated fruit juice for the production of high nutritional beverages. • Clarified kiwi fruit juice had lower viscosity and negligible turbidity because suspended solids were completely removed by ultrafiltration.	Cassano et al., 2007

(Continued)

Table 2.3 (Continued)   Earlier Studies on Hybrid Membrane Process Used in Food Materials

Aim of the Process	Hybrid Membrane Process	Material	Process Condition	Key-results	References
Clarification Concentration	Reverse Osmosis Ultrafiltration	Peach, pear, apple and mandarin juice	RO: 2 and 4 MPa, 25–27°C, 38 L/min. UF: 130 L, 50°C	• It was observed that the use of a pectinolytic enzyme resulted in an approximately 40% increase in permeate flux and preserved the juice properties.	Echavarría et al., 2012
Concentration	Microfiltration Reverse osmosis Nanofiltration	Sea buckthorn	MF: 0.7–3.2 bar, 30°C RO: 9–30 bar, 30°C NF: 9–32 bar, 30°C	• Antioxidant capacity and vitamin C retention was approximately 100% in RO and NF.	Vincze et al., 2007
Concentration	Ultrafiltration Nanofiltration	Milk Whey protein	UF: 30–50°C, 1–5 bar NF: 30–50°C, 0.10–20 bar, 100–200 l/h.	• It was noticed that the suitable temperature of UF is 50°C for milk and whey proteins. • Proper operation parameters of nanofiltration were found as the temperature of 30°C in order to achieve >90% lactose yield.	Atra et al., 2005

treatment. The membranes used in the food industry generally configured as hollow membranes, and the flow occurs as a cross-flow to prevent contamination risk.

Some of the key participants in the food industry membrane technology market are: Aquamarijn Micro Filtration BV, 3M Membranes, Donaldson Co. Inc., Dow Liquid Separations/Filmtec Corp., GE Water Treatment & Process Technologies, Graver Technologies, Koch Membranes Systems Inc., Meissner Filtration Products Inc., Pore technology Inc., Hyflux Ltd., and Toray Industries Inc. Other food companies that are also using membrane technology on an industrial scale are listed below. In a study conducted by Türker et al. (2015), "Pak-Gıda Üretim ve Paz. A.Ş. (Kocaeli/Turkey)" conducted experimental studies with reverse osmosis membrane processes for the separation of naturally occurring rotary filter wastewater from molasses. This membrane system, which was developed as a result of university and industry cooperation, has been implemented in the factory plants since 2014. The wastewater is effectively condensed in the membrane filtration plant utilizing a rotary filter. Moreover, Vinas and Vinas' extract is produced by feeding the remaining usable part to the evaporation plant.

"Gülgün Dairy Products (Nicosia/Cyprus)" uses reverse osmosis techniques for yogurt production to condense raw milk through dehydration. Decomposition occurs based on the molecular size and chemical interactions that occur between the membrane and the milk components. Although high pressure is applied, the milk's proteins are protected. Reverse osmosis preserves the functionality of the proteins and gives the product an uncooked taste. The system is designed and operated to maintain the whole structure of the fat molecules. This prevents quality problems caused by damaged fat molecules (Anonymous, 2019a).

"Porifera Inc. (USA, California)" operates a pilot plant to process 45,000 kg/h milk using forward osmosis systems. The milk is concentrated four times in this system. Forward osmosis provides an energy saving of 44%, steam of 24%, investment cost of 80%, and operating cost of 50% when compared to thermal evaporation. In addition, "ARLA Food (Denmark)" focuses on the application of forward osmosis in dairy products, and they have been researching in order to develop this system.

"Aromsa Besin Aroma ve Katkı Maddeleri Sanayi Ticaret A.Ş. (Kocaeli/Turkey)" uses membrane technology in the aroma sector. They have been applying this technology in the extraction process of vegetables and fruits since 2014. The company prefers membrane technology as this technology does not require heat treatment. When the products are concentrated by membrane technology, aroma and antioxidant capacities

APPLICATIONS OF MEMBRANE TECHNOLOGY

are protected. The liquid extracts produced by membrane technology are encapsulated with suitable carriers and converted into powder form.

"Hazlewood Manor Brewery (Staffordshire, UK)" clarify their beer with an ultrafiltration membrane system. The company has achieved a much cleaner product in less than half the time previously required for clarification, since installing the ultrafiltration plant. Another brewery named "Tower Brewery (Tadcaster, UK)" have designed a 2,000 m³/day capacity reverse osmosis system to increase the daily use of well water. These membrane systems have been supplied by Koch Membrane Systems. Besides these brewery companies, "Ganter (Freiburg, Germany)," "The Moscow Brewing Company (Mytischi, Moscow Region, Russia)," "Princen Brewery (Halfweg, the Netherlands)," "The Joint Stock Company Kamchatskoe Pivo (Kamchatka Kray, Petropavlovsk-Kamchatsky City, Russia)" and "The Lipetskpivo Brewery (Lipetskpivo, Russia)" also uses the membrane filtration system in its beer production. The reasons for choosing this membrane system are to provide a sustainable process, a safe working environment, low costs, constant beer quality, improved taste, and colloidal stability, greater flexibility and modularity, and ease of use (Anonymous, 2019b).

"Coca-Cola FEMSA (Veracruz, Mexico)," one of the franchise bottlers of Coca-Cola Company products, uses the RO facility to treat tap water for use in its products. Coca-Cola FEMSA aims to use this system to standardize water quality. Coca-Cola FEMSA has started using a new membrane cleaning system, and saved 85,000 dollars per year (Anonymous, 2019c).

"Nestlé Waters (Surat Thani, Thailand)" uses reverse osmosis in order to purify water for its Nestlé Pure Life brand. This plant uses a purification system that consists of a primary RO and a VSEP® vibratory RO system. This RO system uses heat-sanitizable membranes and achieves 84% water recovery. Furthermore, the VSEP® vibratory RO system acts like a crystallizer and allows for water recovery even after the brine has been concentrated beyond the solubility limits of the sparingly soluble salt. Furthermore, Nestlé has set up a wastewater treatment and recovery facility to reuse treated water for non-food production applications such as cooling, garden irrigation, and cleaning. Nestlé aims to reduce, reuse, and recycle water in all of its plants. In this treatment plant, an acidification tank, anaerobic digester, ultrafiltration and reverse osmosis slides, a biogas boiler, and ancillary equipment were used. The wastewater to be treated comes from production units, washing stations, and cleaning areas (Anonymous, 2019d, 2019e).

Besides the examples of membrane processes used on an industrial scale, there are many food companies that contribute financially

FREQUENTLY USED MEMBRANE PROCESSING TECHNIQUES

to academic studies. For instance, the "Atlantic Seafood Ingredients Company (La Baule, France)" worked with Cros et al. (2005) on desalinization of mussel (Mytilus edulis) cooking juices with electrodialysis. The researcher investigated the effect of electrodialysis on the aroma profile of mussel cooking juice. United Oil Palm Industry (Nibong tebal, Malaysia) provided samples of palm oil mill effluent (POME) for the research of water recycling from POME using membrane technology that was performed by Ahmad et al., (2003).

2.8 CONCLUSION

The use of membrane systems has been increasing day by day in the food industry. It is possible to achieve selective separation in membrane technologies at low operating temperatures and at low energy demand. Membrane processes, which are defined as a green technology, have less labor requirements, high separation efficiency, and allow for the new process design and optimization of innovative production cycles. However, the high investment costs, concentration polarization, and membrane contamination are the main problems that limit the use of membrane technology. In order to overcome these constraints, mechanisms of the membrane process, driving force, mass transfer, and contamination mechanism should be well evaluated and examined in detail for each process and product. The membrane system has been used in many different sectors in the food industry in order to remove microorganisms from fermented products; remove fat molecules from dairy products; recover valuable ingredients from starch and yeast processing residues; concentrate egg whites, fruit juice and beverages; separate lactose from dairy products; standardize water quality and purify wastewater. The number of membrane systems, namely reverse osmosis, forward osmosis, microfiltration (MF), ultrafiltration (UF), nanofiltration (NF), electrodialysis (ED), membrane distillation, pervaporation (PV), and membrane bioreactor (MBR), are preferred and applied in different processes and product types in the food industry. These systems can be used singularly or also in combination with others. Thus, operating costs can be reduced, and the desired quality product can be obtained while providing the desired separation efficiency. Membrane systems do not cause thermal damage to the product as they are carried out at low operating temperatures. This is why the phase change, denaturation of proteins, and changes in sensory properties of the final product are prevented, and aroma and nutrient loss are reduced, so a high-quality product can be produced. Finally, numerous food manufacturers, such as the Coca-Cola Company, Nestlé water, Pak-Gıda, etc. have used

APPLICATIONS OF MEMBRANE TECHNOLOGY

it in their process and products. Due to the existing advantages provided by membrane technology and technological advances, this technology will find more use in the food industry in the future.

REFERENCES

Aguiar, I.B., Miranda, N.G.M., Gomes, F.S., Santos, M.C.S., Freitas, D., de, G.C., Tonon, R.V., Cabral, L.M.C., 2012. Physicochemical and sensory properties of apple juice concentrated by reverse osmosis and osmotic evaporation. *Innov. Food Sci. Emerg. Technol.* 16, 137–142. doi:10.1016/j.ifset.2012.05.003.

Ahmad, A.L., Ismail, S., Bhatia, S., 2003. Water recycling from palm oil mill effluent (POME) using membrane technology. *Desalination* 157(1–3), 87–95.

Akdağlı, M., Arslan, S., 2008. *Ters Ozmos İleri Arıtma Sistemleri ile Çok Amaçlı Su Arıtımında Proses Tasarımı.* İTÜ Çevre Mühendisliği Bölümü Lisans Tezi, İstanbul, Turkey.

Alvarez, V., Alvarez, S., Riera, F.A., Alvarez, R., 1997. Permeate flux prediction in apple juice concentration by reverse osmosis. *J. Membr. Sci.* 127(1), 25–34. doi:10.1016/S0376-7388(96)00285-2.

Alves, V.D., Coelhoso, I.M., 2006. Orange juice concentration by osmotic evaporation and membrane distillation: A comparative study. *J. Food Eng.* 74(1), 125–133. doi:10.1016/j.jfoodeng.2005.02.019.

Anonymous, 2019a. Yoğurt üretiminde "Membran" teknolojisi. *SUT Dünyası.* URL https://www.sutdunyasi.com/haberler/yogurt-uretiminde-membran-teknolojisi/ (accessed 9.9.19).

Anous, 2019b. Pentair – Beer membrane filtration system – Beverage filtration solutions. *Pentair Food Beverage Process. Solut.* URL https://foodandbeverage.pentair.com/en/products/beverage-filtration-solutions-bmf-18 (accessed 9.9.19).

Anous, 2019c. Coca-Cola FEMSA, reverse osmosis plant. URL https://www.pwtchemicals.com/resources/coca-cola-femsa/ (accessed 9.9.19).

Anous, 2019d. Food industry: Nestlé chooses veolia for recycling water | Veolia. URL https://www.veolia.com/en/recycling-water-food-industry-nestle (accessed 9.9.19).

Anous, 2019e. Nestlé bottling plant goes ZLD with VSEP – New logic research URL https://www.vsep.com/articles/34/ (accessed 9.9.19).

Arriola, N.A., dos Santos, G.D., Prudêncio, E.S., Vitali, L., Petrus, J.C.C., Castanho Amboni, R.D., 2014. Potential of nanofiltration for the concentration of bioactive compounds from watermelon juice. *Int. J. Food Sci. Technol.* 49(9), 2052–2060.

Artiga, P., Ficara, E., Malpei, F., Garrido, J.M., Méndez, R., 2005. Treatment of two industrial wastewaters in a submerged membrane bioreactor. *Desalin. Membr. Drinking Ind. Water Prod.* 179(1–3), 161–169. doi:10.1016/j.desal.2004.11.064.

Aslan, M., 2016. *Membran Teknolojileri.* TC çevre ve Şehircilik Bakanl. Türkiye çevre koruma Vakfı, Ankara, Turkey.

Atra, R., Vatai, G., Bekassy-Molnar, E., Balint, A., 2005. Investigation of ultra- and nanofiltration for utilization of whey protein and lactose. *J. Food Eng.* 67(3), 325–332. doi:10.1016/j.jfoodeng.2004.04.035.

Babu, B.R., Rastogi, N.K., Raghavarao, K.S.M.S., 2006. Effect of process parameters on transmembrane flux during direct osmosis. *J. Membr. Sci.* 280(1–2), 185–194. doi:10.1016/j.memsci.2006.01.018.

Baker, R.W., Cussler, E.L., Eykamp, W., Koros, W.J., Riley, R.L., Strathmann, H., 1990. *Membrane Separation Systems – A Research and Development Needs Assessment. USDOE.* Office of Energy Research, Washington, DC.

Banat, F.A., Simandl, J., 1996. Removal of benzene traces from contaminated water by vacuum membrane distillation. *Chem. Eng. Sci.* 51(8), 1257–1265. doi:10.1016/0009-2509(95)00365-7.

Bandini, S., Sarti, G.C., 2002. Concentration of must through vacuum membrane distillation. *Desalination* 149(1–3), 253–259. doi:10.1016/S0011-9164(02)00776-2.

Bánvölgyi, S., Horváth, S., Stefanovits-Bányai, É., Békássy-Molnár, E., Vatai, G., 2009. Integrated membrane process for blackcurrant (Ribes nigrum L.) juice concentration. *Desalination, the Third Membrane Science and Technology Conference of Visegrad Countries (PERMEA); Part 2 241*, pp. 281–287. doi:10.1016/j.desal.2007.11.088.

Barlas, H., 2002. *Suların arıtımında ileri teknolojiler" ders notları.* İÜ Mühendis. Fakültesi Çevre Mühendisliği Bölümü Avcılar, İstanbul, Turkey.

Beaudry, E.G., Lampi, K.A., 1990. Membrane technology for direct osmosis concentration of fruit juices. *Food Technol.* 6, 121.

Beolchini, F., Vegho', F., Barba, D., 2004. Microfiltration of bovine and ovine milk for the reduction of microbial content in a tubular membrane: a preliminary investigation, Vegho', F., Barba, D. *Desalination* 161(3), 251–258. doi:10.1016/S0011-9164(03)00705-7.

Bernardes, A., Rodrigues, M., Ferreira, J.Z., 2016. *Electrodialysis and Water Reuse.* Springer.

Bhattacharjee, C., Saxena, V.K., Dutta, S., 2017. Fruit juice processing using membrane technology: A review. *Innov. Food Sci. Emerg. Technol.* 43, 136–153.

Brazinha, C., Crespo, J.G., 2009. Aroma recovery from hydro alcoholic solutions by organophilic pervaporation: Modelling of fractionation by condensation. *J. Membr. Sci.* 341(1–2), 109–121. doi:10.1016/j.memsci.2009.05.045.

Camacho, L., Dumée, L., Zhang, J., Li, J., Duke, M., Gomez, J., Gray, S., 2013. Advances in membrane distillation for water desalination and purification applications. *Water* 5(1), 94–196.

Cardew, P.T., 2007. *Membrane Processes: A Technology Guide.* Royal Society of Chemistry.

Carneiro, L., dos Santos Sa, I., dos Santos Gomes, F., Matta, V.M., Cabral, L.M.C., 2002. Cold sterilization and clarification of pineapple juice by tangential microfiltration. *Desalination* 148(1–3), 93–98. doi:10.1016/S0011-9164(02)00659-8.

Cassano, A., Conidi, C., Drioli, E., 2011. Comparison of the performance of UF membranes in olive mill wastewater treatment. *Water Res.* 45(10), 3197–3204.

Cassano, A., Donato, L., Drioli, E., 2007. Ultrafiltration of kiwifruit juice: Operating parameters, juice quality and membrane fouling. *J. Food Eng.* 79(2), 613–621. doi:10.1016/j.jfoodeng.2006.02.020.

APPLICATIONS OF MEMBRANE TECHNOLOGY

Cassano, A., Drioli, E., 2013. *Integrated Membrane Operations: In the Food Production.* Walter de Gruyter, Berlin, Germany.

Cassano, A., Molinari, R., Romano, M., Drioli, E., 2001. Treatment of aqueous effluents of the leather industry by membrane processes: A review. *J. Membr. Sci.* 181(1), 111–126. doi:10.1016/S0376-7388(00)00399-9.

Catarino, M., Ferreira, A., Mendes, A., 2009. Study and optimization of aroma recovery from beer by pervaporation. *J. Membr. Sci.* 341(1–2), 51–59. doi:10.1016/j.memsci.2009.05.038.

Catarino, M., Mendes, A., 2011. Dealcoholizing wine by membrane separation processes. *Innov. Food Sci. Emerg. Technol.* 12(3), 330–337. doi:10.1016/j.ifset.2011.03.006.

Cath, T.Y., Childress, A.E., Elimelech, M., 2006. Forward osmosis: Principles, applications, and recent developments. *J. Membr. Sci.* 281(1–2), 70–87.

Cath, T.Y., Gormly, S., Beaudry, E.G., Flynn, M.T., Adams, V.D., Childress, A.E., 2005. Membrane contactor processes for wastewater reclamation in space: Part I. Direct osmotic concentration as pretreatment for reverse osmosis. *J. Membr. Sci.* 257(1–2), 85–98.

Charcosset, C., 2006. Membrane processes in biotechnology: An overview. *Biotechnol. Adv.* 24(5), 482–492. doi:10.1016/j.biotechadv.2006.03.002.

Cheng, S., Oatley, D.L., Williams, P.M., Wright, C.J., 2011. Positively charged nanofiltration membranes: Review of current fabrication methods and introduction of a novel approach. *Adv. Colloid Interface Sci.* 164(1–2), 12–20.

Cheremisinoff, N.P., 2002. *Handbook of Water and Wastewater Treatment Technologies.* Butterworth-Heinemann, Boston, MA.

Cheryan, M., 1998. *Ultrafiltration and Microfiltration Handbook.* CRC Press, Boca Raton, FL.

Christensen, K., Andresen, R., Tandskov, I., Norddahl, B., du Preez, J.H., 2006. Using direct contact membrane distillation for whey protein concentration. *Desalination* 200(1–3), 523–525. doi:10.1016/j.desal.2006.03.421.

Chung, T.-S., Zhang, S., Wang, K.Y., Su, J., Ling, M.M., 2012. Forward osmosis processes: Yesterday, today and tomorrow. *Desalination* 287, 78–81.

Çinar, Ö., Hasar, H., Kinaci, C., 2006. Modeling of submerged membrane bioreactor treating cheese whey wastewater by artificial neural network. *J. Biotechnol.* 123(2), 204–209. doi:10.1016/j.jbiotec.2005.11.002.

Conidi, C., Cassano, A., Drioli, E., 2011. A membrane-based study for the recovery of polyphenols from bergamot juice. *J. Membr. Sci.* 375(1–2), 182–190. doi:10.1016/j.memsci.2011.03.035.

Crespo, J.G., Böddeker, K.W., 2013. *Membrane Processes in Separation and Purification.* Springer Science & Business Media.

Cros, S., Lignot, B., Bourseau, P., Jaouen, P., Prost, C., 2005. Desalination of mussel cooking juices by electrodialysis: Effect on the aroma profile. *J. Food Eng.* 69(4), 425–436.

Curcio, E., Drioli, E., 2005. Membrane distillation and related operations – A review. *Sep. Purif. Rev.* 34(1), 35–86.

de Bruijn, J.P.F., Venegas, A., Martínez, J.A., Bórquez, R., 2003. Ultrafiltration performance of Carbosep membranes for the clarification of apple juice. *LWT Food Sci. Technol.* 36(4), 397–406. doi:10.1016/S0023-6438(03)00015-X.

de Morais Coutinho, C., Chiu, M.C., Basso, R.C., Ribeiro, A.P.B., Gonçalves, L.A.G., Viotto, L.A., 2009. State of art of the application of membrane technology to vegetable oils: A review. *Food Res. Int.* 42(5–6), 536–550.

Diban, N., Voinea, O.C., Urtiaga, A., Ortiz, I., 2009. Vacuum membrane distillation of the main pear aroma compound: Experimental study and mass transfer modeling. *J. Membr. Sci.* 326(1), 64–75.

Dizge, N., 2011. Mikrofiltrasyon membranların kirlenme özelliklerinin Membran tipine ve gözenek boyutuna bağlı olarak klasik aktif çamur sisteminde incelenmesi (PhD Thesis). Doktora Tezi, Gebze Yüksek Teknoloji Enstitüsü Mühendislik ve Fen Bilimleri.

Drioli, E., Fontananova, E., 2004. Membrane technology and sustainable growth. *Chem. Eng. Res. Des.* 82(12), 1557–1562.

Echavarría, A.P., Falguera, V., Torras, C., Berdún, C., Pagán, J., Ibarz, A., 2012. Ultrafiltration and reverse osmosis for clarification and concentration of fruit juices at pilot plant scale. *LWT Food Sci. Technol.* 46(1), 189–195. doi:10.1016/j.lwt.2011.10.008.

Echavarría, A.P., Torras, C., Pagán, J., Ibarz, A., 2011. Fruit juice processing and membrane technology application. *Food Eng. Rev.* 3(3–4), 136–158.

El-Abbassi, A., Kiai, H., Raiti, J., Hafidi, A., 2014. Application of ultrafiltration for olive processing wastewater treatment. *J. Clean. Prod.* 65, 432–438. doi:10.1016/j.jclepro.2013.08.016.

El-Sayed, M.M.H., Chase, H.A., 2011. Trends in whey protein fractionation. *Biotechnol. Lett.* 33(8), 1501–1511. doi:10.1007/s10529-011-0594-8.

Farid, M.M., 2010. *Mathematical Modeling of Food Processing.* CRC Press, Boca Raton, FL.

Feng, X., Huang, R.Y.M., 1997. Liquid separation by membrane pervaporation: A review. *Ind. Eng. Chem. Res.* 36(4), 1048–1066. doi:10.1021/ie960189g.

Franken, A.C.M., Mulder, M.H.V., Smolders, C.A., 1990. Pervaporation process using a thermal gradient as the driving force. *J. Membr. Sci.* 53(1–2), 127–141. doi:10.1016/0376-7388(90)80009-B.

Garcia-Castello, E., Cassano, A., Criscuoli, A., Conidi, C., Drioli, E., 2010. Recovery and concentration of polyphenols from olive mill wastewaters by integrated membrane system. *Water Res.* 44(13), 3883–3892. doi:10.1016/j.watres.2010.05.005.

García-Martín, N., Perez-Magariño, S., Ortega-Heras, M., González-Huerta, C., Mihnea, M., González-Sanjosé, M.L., Palacio, L., Prádanos, P., Hernández, A., 2010. Sugar reduction in musts with nanofiltration membranes to obtain low alcohol-content wines. *Sep. Purif. Technol.* 76(2), 158–170.

Giwa, A., Hasan, S.W., 2015. Numerical modeling of an electrically enhanced membrane bioreactor (MBER) treating medium-strength wastewater. *J. Environ. Manag.* 164, 1–9. doi:10.1016/j.jenvman.2015.08.031.

Goh, P.S., Matsuura, T., Ismail, A.F., Hilal, N., 2016. Recent trends in membranes and membrane processes for desalination. *Desalination* 391, 43–60.

APPLICATIONS OF MEMBRANE TECHNOLOGY

Gómez Benítez, J., Palacios Macías, V.M., Szekely Gorostiaga, P., Veas López, R., Pérez Rodríguez, L., 2003. Comparison of electrodialysis and cold treatment on an industrial scale for tartrate stabilization of sherry wines. *J. Food Eng.* 58(4), 373–378. doi:10.1016/S0260-8774(02)00421-1.

Gonçalves, F., Fernandes, C., Cameira dos Santos, P., de Pinho, M.N., 2003. Wine tartaric stabilization by electrodialysis and its assessment by the saturation temperature. *J. Food Eng.* 59(2–3), 229–235. doi:10.1016/S0260-8774(02)00462-4.

Gonçalves, M.D.C., Windmöller, D., Erismann, N.de M., Galembeck, F., 1990. Pressure-driven pervaporation. *Sep. Sci. Technol.* 25(9–10), 1079–1085. doi:10.1080/01496399008050386.

Grebenyuk, V.D., Grebenyuk, O.V., 2002. Electrodialysis: from an idea to realization. *Russ. J. Electrochem.* 38(8), 806–809.

Gudernatsch, W., Kimmerle, K., 1991. Pilot dealcoholization of beer by pervaporation. *Proc Fifth International Conference Pervapor Process. Chem Indus*, pp. 291–307.

Gurak, P.D., Cabral, L.M., Rocha-Leão, M.H.M., Matta, V.M., Freitas, S.P., 2010. Quality evaluation of grape juice concentrated by reverse osmosis. *J. Food Eng.* 96(3), 421–426.

Gürel, L., Büyükgüngör, H., 2015. Kütle Aktarımının Membran Sistemlerindeki rolü. *Pamukkale Univ. J. Eng. Sci.* 21(6): 224–238.

Gyura, J., Šereš, Z., Vatai, G., Molnár, E.B., 2002. Separation of non-sucrose compounds from the syrup of sugar-beet processing by ultra-and nanofiltration using polymer membranes. *Desalination* 148(1–3), 49–56.

Hacıfettahoğlu, A., 2009. Biyodizel üretim Tesisi Atıksularının Membran Filtrasyonu (MSc Thesis). Yüksek Lisans Tezi, Kocaeli Üniversitesi Fen Bilimleri Enstitüsü, Kocaeli.

Hasanoğlu, A., Gül, K., 2016. Concentration of skim milk and dairy products by forward osmosis. *J. Turk. Chem. Soc. Sect. B Chem. Eng.* 1, 149–160.

Hasar, H., 2001. Batık membran-aktif çamur sistemlerinin arıtma kapasitesinin geliştirilmesi ve modellenmesi/Development of treatment capacity of submerged membrane activated sludge systems and its modelling.

Hausmann, A., Sanciolo, P., Vasiljevic, T., Ponnampalam, E., Quispe-Chavez, N., Weeks, M., Duke, M., 2011. Direct contact membrane distillation of dairy process streams. *Membranes* 1(1), 48–58. doi:10.3390/membranes1010048.

Hedrick, T.I., 1983. Reverse osmosis and ultrafiltration in the food industry. *Dry. Technol.* 2, 329–352.

Ho, W., Sirkar, K., 2012. *Membrane Handbook*. Springer Science & Business Media.

Hoinkis, J., Deowan, S.A., Panten, V., Figoli, A., Huang, R.R., Drioli, E., 2012. Membrane bioreactor (MBR) technology – A promising approach for industrial water reuse. *Procedia Eng. SWEE* 11(33), 234–241. doi:10.1016/j.proeng.2012.01.1199.

Huang, R.Y.M., 1991. Pervaporation membrane separation processes. *Membr. Sci. Technol. Ser.* 1, 111.

Hussain, A.A., Al-Rawajfeh, A.E., 2009. Recent patents of nanofiltration applications in oil processing, desalination, wastewater and food industries. *Recent Pat. Chem. Eng.* 2(1), 51–66.

Jiao, B., Cassano, A., Drioli, E., 2004. Recent advances on membrane processes for the concentration of fruit juices: A review. *J. Food Eng.* 63(3), 303–324. doi:10.1016/j.jfoodeng.2003.08.003.

Jiraratananon, R., Chanachai, A., Huang, R.Y.M., Uttapap, D., 2002. Pervaporation dehydration of ethanol–water mixtures with chitosan/hydroxyethylcellulose (CS/HEC) composite membranes: I. Effect of operating conditions. *J. Membr. Sci.* 195(2), 143–151. doi:10.1016/S0376-7388(01)00563-4.

Juda, W., McRae, W.A., 1950. Coherent ion-exchange gels and membranes. *J. Am. Chem. Soc.* 72(2), 1044–1044.

Judd, S., 2010. *The MBR Book: Principles and Applications of Membrane Bioreactors for Water and Wastewater Treatment.* Elsevier.

Kang, G., Cao, Y., 2012. Development of antifouling reverse osmosis membranes for water treatment: A review. *Water Res.* 46(3), 584–600.

Karlsson, H.O.E., Loureiro, S., Trägårdh, G., 1995. Aroma compound recovery with pervaporation – Temperature effects during pervaporation of a Muscat wine. *J. Food Eng.* 26(2), 177–191. doi:10.1016/0260-8774(94)00050-J.

Kober, P.A., 1917. Pervaporation, perstillation and percrystallization. *J Am. Chem. Soc.* 39(5), 944–948.

Koçak, İ., 2007. Ters osmoz sistemiyle sudan borun uzaklaştırılması (PhD Thesis). Selçuk Üniversitesi Fen Bilimleri Enstitüsü.

Kotsanopoulos, K.V., Arvanitoyannis, I.S., 2015. Membrane processing technology in the food industry: Food processing, wastewater treatment, and effects on physical, microbiological, organoleptic, and nutritional properties of foods. *Crit. Rev. Food Sci. Nutr.* 55(9), 1147–1175.

Koyuncu, İ., 2001. Nanofiltrasyon membrantları ile tuz gideriminde organik iyon etkisi (PhD Thesis). Fen Bilimleri Enstitüsü.

Koyuncu, İ., 2018. *Su/Atıksu Arıtılması ve geri Kazanılmasında Membran Teknolojileri ve Uygulamaları*, 1st ed. Yıldızlar Ofset, Ankara, Turkey.

Kozák, Á., Rektor, A., Vatai, G., 2006. Integrated large-scale membrane process for producing concentrated fruit juices. *Desalination* 200(1–3), 540–542. doi:10.1016/j.desal.2006.03.428.

Kumar, P., Sharma, N., Ranjan, R., Kumar, S., Bhat, Z.F., Jeong, D.K., 2013. Perspective of membrane technology in dairy industry: A review. *Asian-Australas. J. Anim. Sci.* 26(9), 1347.

Kuruma, H., Poetzschke, J., 2002. İçme sularında amonyum iyonlarının uzaklaştırılmasında Membran filtrasyon uygulaması. *Ekoloji* 11, 45–48.

Kwak, R., Guan, G., Peng, W.K., Han, J., 2013. Microscale electrodialysis: Concentration profiling and vortex visualization. *Desalination* 308, 138–146.

Laganà, F., Barbieri, G., Drioli, E., 2000. Direct contact membrane distillation: Modelling and concentration experiments. *J. Membr. Sci.* 166(1), 1–11. doi:10.1016/S0376-7388(99)00234-3.

Le-Clech, P., Chen, V., Fane, T.A.G., 2006. Fouling in membrane bioreactors used in wastewater treatment. *J. Membr. Sci.* 284(1–2), 17–53. doi:10.1016/j.memsci.2006.08.019.

Lee, K.P., Arnot, T.C., Mattia, D., 2011. A review of reverse osmosis membrane materials for desalination – Development to date and future potential. *J. Membr. Sci.* 370(1–2), 1–22.

Lesjean, B., Huisjes, E.H., 2008. Survey of the European MBR market: Trends and perspectives. *Desalination, Selected Papers Presented at the 4th International IWA Conference on Membranes for Water and Wastewater Treatment*, May 15–17, 2007, Harrogate, UK. Guest Edited by Simon Judd; and Papers Presented at the International Workshop on Membranes and Solid-Liquid Separation Processes, July 11, 2007, INSA, Toulouse, France. Guest edited by Saravanamuthu Vigneswaran and Jaya Kandasamy 231, 71–81. doi:10.1016/j.desal.2007.10.022.

Li, D., Wang, H., 2010. Recent developments in reverse osmosis desalination membranes. *J. Mater. Chem.* 20(22), 4551–4566.

Lutchmiah, K., Verliefde, A.R.D., Roest, K., Rietveld, L.C., Cornelissen, E.R., 2014. Forward osmosis for application in wastewater treatment: A review. *Water Res.* 58, 179–197.

Malaeb, L., Ayoub, G.M., 2011. Reverse osmosis technology for water treatment: State of the art review. *Desalination* 267(1), 1–8.

Marella, C., Muthukumarappan, K., Metzger, L.E., 2013. Application of membrane separation technology for developing novel dairy food ingredients. *J. Food Process. Technol.* 4, 10–4172.

Massot, A., Mietton-Peuchot, M., Peuchot, C., Milisic, V., 2008. Nanofiltration and reverse osmosis in winemaking. *Desalination* 231(1–3), 283–289.

Matin, A., Khan, Z., Zaidi, S.M.J., Boyce, M.C., 2011. Biofouling in reverse osmosis membranes for seawater desalination: Phenomena and prevention. *Desalination* 281, 1–16.

McGinnis, R.L., Elimelech, M., 2007. Energy requirements of ammonia–carbon dioxide forward osmosis desalination. *Desalination* 207(1–3), 370–382.

Merin, U., 1986. Bacteriological aspects of microfiltration of cheese whey. *J. Dairy Sci.* 69(2), 326–328.

Mi, B., Elimelech, M., 2008. Chemical and physical aspects of organic fouling of forward osmosis membranes. *J. Membr. Sci.* 320(1–2), 292–302.

Mi, B., Elimelech, M., 2010. Organic fouling of forward osmosis membranes: Fouling reversibility and cleaning without chemical reagents. *J. Membr. Sci.* 348(1–2), 337–345.

Misdan, N., Lau, W.J., Ismail, A.F., 2012. Seawater Reverse Osmosis (SWRO) desalination by thin-film composite membrane – Current development, challenges and future prospects. *Desalination* 287, 228–237.

Miyaki, H., Adachi, S., Suda, K., Kojima, Y., 2000. Water recycling by floating media filtration and nanofiltration at a soft drink factory. *Desalination* 131(1–3), 47–53.

Moresi, M., Sappino, F., 1998. Economic feasibility study of citrate recovery by electrodialysis. *J. Food Eng.* 35(1), 75–90. doi:10.1016/S0260-8774(98)00012-0.

Mulder, J., 2012. *Basic Principles of Membrane Technology.* Springer Science & Business Media.

Nayak, C.A., Rastogi, N.K., 2010. Comparison of osmotic membrane distillation and forward osmosis membrane processes for concentration of anthocyanin. *Desalin. Water Treat.* 16(1–3), 134–145. doi:10.5004/dwt.2010.1084.

Nayak, C.A., Valluri, S.S., Rastogi, N.K., 2011. Effect of high or low molecular weight of components of feed on transmembrane flux during forward osmosis. *J. Food Eng.* 106(1), 48–52. doi:10.1016/j.jfoodeng.2011.04.006.

Nguyen, T.P.N., Yun, E.-T., Kim, I.-C., Kwon, Y.-N., 2013. Preparation of cellulose triacetate/cellulose acetate (CTA/CA)-based membranes for forward osmosis. *J. Membr. Sci.* 433, 49–59.

Noble, R.D., Stern, S.A., 1995. *Membrane Separations Technology: Principles and Applications.* Elsevier.

Nunes, S.P., Peinemann, K.-V., 2001. *Membrane Technology.* Wiley Online Library, Weinheim.

Ochando-Pulido, J.M., Verardo, V., Segura-Carretero, A., Martinez-Ferez, A., 2015. Analysis of the concentration polarization and fouling dynamic resistances under reverse osmosis membrane treatment of olive mill wastewater. *J. Ind. Eng. Chem.* 31, 132–141.

Olmo, Á. del, Blanco, C.A., Palacio, L., Prádanos, P., Hernández, A., 2014. Pervaporation methodology for improving alcohol-free beer quality through aroma recovery. *J. Food Eng.* 133, 1–8. doi:10.1016/j.jfoodeng.2014.02.014.

Onsekizoglu, P., Bahceci, K.S., Acar, M.J., 2010. Clarification and the concentration of apple juice using membrane processes: A comparative quality assessment. *J. Membr. Sci.* 352(1–2), 160–165.

Pacheco, M.A., Marshall, C.L., 1997. Review of dimethyl carbonate (DMC) manufacture and its characteristics as a fuel additive. *Energy Fuels* 11(1), 2–29. doi:10.1021/ef9600974.

Pak, Ü., 2011. Ekmek Mayası Endüstrisi Seperasyon Prosesi Atıksularında Membran Prosesler Ile Renk Giderimi (PhD Thesis). Fen Bilimleri Enstitüsü.

Palacios, V.M., Caro, I., Pérez, L., 2002. Comparative study of crossflow microfiltration with conventional filtration of sherry wines. *J. Food Eng.* 54(2), 95–102. doi:10.1016/S0260-8774(01)00189-3.

Pan, B., Yan, P., Zhu, L., Li, X., 2013. Concentration of coffee extract using nanofiltration membranes. *Desalination* 317, 127–131. doi:10.1016/j.desal.2013.03.004.

Pap, N., Kertész, Sz., Pongrácz, E., Myllykoski, L., Keiski, R.L., Vatai, Gy., László, Zs., Beszédes, S., Hodúr, C., 2009. Concentration of blackcurrant juice by reverse osmosis. *Desalination, The Third Membrane Science and Technology Conference of Visegrad Countries (PERMEA); Part 2* 241, pp. 256–264. doi:10.1016/j.desal.2008.01.069.

Paredes Valencia, A., Doyen, A., Benoit, S., Margni, M., Pouliot, Y., 2018. Effect of ultrafiltration of milk prior to fermentation on mass balance and process efficiency in Greek-style yogurt manufacture. *Foods* 7(9), 144. doi:10.3390/foods7090144.

Pederson, C.L., Lueptow, R.M., 2007. Fouling in a high pressure, high recovery rotating reverse osmosis system. *Desalination* 212(1–3), 1–14.

Pilipovik, M.V., Riverol, C., 2005. Assessing dealcoholization systems based on reverse osmosis. *J. Food Eng.* 69(4), 437–441. doi:10.1016/j.jfoodeng.2004.08.035.

APPLICATIONS OF MEMBRANE TECHNOLOGY

Pouliot, Y., 2008. Membrane processes in dairy technology – From a simple idea to worldwide panacea. *Int. Dairy J.* 18(7), 735–740.

Qasim, M., Badrelzaman, M., Darwish, N.N., Darwish, N.A., Hilal, N., 2019. Reverse osmosis desalination: A state-of-the-art review. *Desalination* 459, 59–104.

Qiu, N., Tian, Y., Qiao, S., Deng, H., 2009. Apple pectin behavior separated by ultra-filtration. *Agric. Sci. China* 8(10), 1193–1202. doi:10.1016/S1671-2927(08)60329-6.

Ramírez, C.A., Patel, M., Blok, K., 2006. From fluid milk to milk powder: Energy use and energy efficiency in the European dairy industry. *Energy* 31(12), 1984–2004. doi:10.1016/j.energy.2005.10.014.

Rautenbach, R.M., 1997. *Grundlagen der Modul-und Anlagenauslegung.* Springer.

Ripperger, S., Altmann, J., 2002. Crossflow microfiltration–state of the art. *Sep. Purif. Technol.* 26(1), 19–31.

Rodrigues, S., Fernandes, F.A.N., 2012. *Advances in Fruit Processing Technologies.* CRC Press: Boca Raton, FL.

Rossignol, N., Vandanjon, L., Jaouen, P., Quemeneur, F., 1999. Membrane technology for the continuous separation microalgae/culture medium: Compared performances of cross-flow microfiltration and ultrafiltration. *Aquac Eng.* 20(3), 191–208.

Salt, Y., Dinçer, S., Atmaca, M., İnan, A.T., Gül, M.Z., Adıgüzelov, E., Sezer, Y., Akgün, N., Akgün, M., Kaykıoğlu, G., 2006. Özel Ayırma İşlemlerinde Bir Seçenek: Membran Prosesleri. *Sigma J. Eng. Nat. Sci.* 24(4), 1–23.

Saltık, M.B., Özkan, L., Jacobs, M., van der Padt, A., 2017. Dynamic modeling of ultrafiltration membranes for whey separation processes. *Comput. Chem. Eng.* 99, 280–295. doi:10.1016/j.compchemeng.2017.01.035.

Semmens, M.J., Qin, R., Zander, A., 1989. Using a microporous hollow-fiber membrane to separate VOCs From water. *J. AWWA* 81, 162–167. doi:10.1002/j.1551-8833.1989.tb03195.x.

Shah, D., Bhattacharyya, D., Ghorpade, A., Mangum, W., 1999. Pervaporation of pharmaceutical waste streams and synthetic mixtures using water selective membranes. *Environ. Prog.* 18(1), 21–29. doi:10.1002/ep.670180116.

Singh, R., 1998. Industrial membrane separation processes. *Chemtech*, 28(4), 33–44.

Skilhagen, S.E., Dugstad, J.E., Aaberg, R.J., 2008. Osmotic power – Power production based on the osmotic pressure difference between waters with varying salt gradients. *Desalination* 220(1–3), 476–482.

Smitha, B., Suhanya, D., Sridhar, S., Ramakrishna, M., 2004. Separation of organic–organic mixtures by pervaporation – A review. *J. Membr. Sci.* 241(1), 1–21. doi:10.1016/j.memsci.2004.03.042.

Strathmann, H., 2001. Membrane separation processes: Current relevance and future opportunities. *AIChE J.* 47(5), 1077–1087.

Strathmann, H., 2010. Electrodialysis, a mature technology with a multitude of new applications. *Desalination* 264(3), 268–288.

Strathmann, H., Giorno, L., Drioli, E., 2011. *Introduction to Membrane Science and Technology.* Wiley-VCH, Weinheim.

Takács, L., Vatai, G., Korány, K., 2007. Production of alcohol free wine by pervaporation. *J. Food Eng.* 78(1), 118–125. doi:10.1016/j.jfoodeng.2005.09.005.

Takht Ravanchi, M., Kaghazchi, T., Kargari, A., 2009. Application of membrane separation processes in petrochemical industry: A review. *Desalination* 235(1–3), 199–244. doi:10.1016/j.desal.2007.10.042.

Tansel, B., Sager, J., Rector, T., Garland, J., Strayer, R.F., Levine, L., Roberts, M., Hummerick, M., Bauer, J., 2006. Significance of hydrated radius and hydration shells on ionic permeability during nanofiltration in dead end and cross flow modes. *Sep. Purif. Technol.* 51(1), 40–47. doi:10.1016/j.seppur.2005.12.020.

Taşıyıcı, S., 2009. Batık Membran Sistemleri Ile İçme Suyu Arıtımı: Membran Tıkanıklığını Azaltmak İçin Farklı Yöntemlerin Kullanılması (PhD Thesis). Fen Bilimleri Enstitüsü.

Teorell, T., 1935. Studies on the "Diffusion Effect" upon ionic distribution. Some theoretical considerations. *Proc. Natl. Acad. Sci. U. S. A.* 21(3), 152.

Thorsen, T., Holt, T., 2009. The potential for power production from salinity gradients by pressure retarded osmosis. *J. Membr. Sci.* 335(1–2), 103–110.

Tomaszewska, M., 1993. Concentration of the extraction fluid from sulfuric acid treatment of phosphogypsum by membrane distillation. *J. Membr. Sci.* 78(3), 277–282. doi:10.1016/0376-7388(93)80007-K.

Tomaszewska, M., Gryta, M., Morawski, A.W., 2001. Recovery of hydrochloric acid from metal pickling solutions by membrane distillation. *Sep. Purif. Technol.*, 22–23, 591–600. doi:10.1016/S1383-5866(00)00164-7.

Turano, E., Curcio, S., De Paola, M.G., Calabrò, V., Iorio, G., 2002. An integrated centrifugation–ultrafiltration system in the treatment of olive mill wastewater. *J. Membr. Sci.* 209(2), 519–531.

Türker, M., Karadağ, S., Işık, Y., Ertan, İ., 2015. Maya Endüstrisi 1Koku problemi ve Çözümleri: PAKMAYA deneyimi. *Presented at the 6. Ulus. Hava Kirliliği Kontrolü Sempozyumu İzmir Turkey.*

Van Den Berg, J.C.T., 1962. Evaporated and condensed milk. *Milk Hyg. Monogr. WHO Geneva Switz. WHO Monogr. Ser.*, 48, 321–345.

Varming, C., Andersen, M.L., Poll, L., 2004. Influence of thermal treatment on black currant (Ribes nigrum L.) juice aroma. *J. Agric. Food Chem.* 52(25), 7628–7636. doi:10.1021/jf049435m.

Verhoef, A., Figoli, A., Leen, B., Bettens, B., Drioli, E., Van der Bruggen, B., 2008. Performance of a nanofiltration membrane for removal of ethanol from aqueous solutions by pervaporation. *Sep. Purif. Technol.* 60(1), 54–63. doi:10.1016/j.seppur.2007.07.044.

Versari, A., Ferrarini, R., Parpinello, G.P., Galassi, S., 2003. Concentration of grape must by nanofiltration membranes. *Food Bioprod. Process. Cereal Process.* 81(3), 275–278. doi:10.1205/096030803322438045.

Vigneswaran, S., Ngo, H.H., Chaudhary, D.S., Hung, Y.-T., 2005. Physicochemical treatment processes for water reuse. In: Wang, L.K., Hung, Y.-T., Shammas, N.K. (Eds.), *Physicochemical Treatment Processes, Handbook of Environmental Engineering*, Humana Press, Totowa, NJ, pp. 635–676. doi:10.1385/1-59259-820-x:635.

Vincze, I., Bányai-Stefanovits, É., Vatai, G., 2007. Concentration of sea buckthorn (Hippophae rhamnoides L.) juice with membrane separation.

Sep. Purif. Technol. 455–460. PERMEA 2005 Special Issue 57. doi:10.1016/j.seppur.2006.06.020.

Vivekanand, V., Iyer, M., Ajlouni, S., 2012. Clarification and stability enhancement of pear juice using loose nanofiltration. *J. Food Process. Technol.* 3, 1–6.

Wallace, M., Cui, Z., Hankins, N.P., 2008. A thermodynamic benchmark for assessing an emergency drinking water device based on forward osmosis. *Desalination* 227(1–3), 34–45.

Wang, P., Chung, T.-S., 2015. Recent advances in membrane distillation processes: Membrane development, configuration design and application exploring. *J. Membr. Sci.* 474, 39–56.

Warczok, J., Ferrando, M., Lopez, F., Güell, C., 2004. Concentration of apple and pear juices by nanofiltration at low pressures. *J. Food Eng.* 63(1), 63–70.

Williams, M.E., 2003. *A Brief Review of Reverse Osmosis Membrane Technology.* EET Corp. Williams Eng. Serv. Co., pp. 1–29.

Xu, J., Zhang, L., Gao, X., Bie, H., Fu, Y., Gao, C., 2015. Constructing antimicrobial membrane surfaces with polycation–copper (II) complex assembly for efficient seawater softening treatment. *J. Membr. Sci.* 491, 28–36.

Yazdanshenas, M., Tabatabaee-Nezhad, S.A.R., Soltanieh, M., Roostaazad, R., Khoshfetrat, A.B., 2010. Contribution of fouling and gel polarization during ultrafiltration of raw apple juice at industrial scale. *Desalination* 258(1–3), 194–200. doi:10.1016/j.desal.2010.03.014.

Yip, N.Y., Tiraferri, A., Phillip, W.A., Schiffman, J.D., Elimelech, M., 2010. High performance thin-film composite forward osmosis membrane. *Environ. Sci. Technol.* 44(10), 3812–3818.

Yuan, W., Zydney, A.L., 1999. Humic acid fouling during microfiltration. *J. Membr. Sci.* 157(1), 1–12.

Zakrzewska-Trznadel, G., Harasimowicz, M., Chmielewski, A.G., 1999. Concentration of radioactive components in liquid low-level radioactive waste by membrane distillation. *J. Membr. Sci.* 163(2), 257–264. doi:10.1016/S0376-7388(99)00171-4.

Zhao, L., Ho, W.W., 2014. Novel reverse osmosis membranes incorporated with a hydrophilic additive for seawater desalination. *J. Membr. Sci.* 455, 44–54.

Zhou, H., Smith, D.W., 2002. Advanced technologies in water and wastewater treatment. *J. Environ. Eng. Sci.* 1(4), 247–264. doi:10.1139/s02-020.

Zirehpour, A., Rahimpour, A., Jahanshahi, M., Peyravi, M., 2014. Mixed matrix membrane application for olive oil wastewater treatment: Process optimization based on Taguchi design method. *J. Environ. Manag.* 132, 113–120.

Zolotarev, P.P., Ugrozov, V.V., Volkina, I.B., Nikulin, V.M., 1994. Treatment of waste water for removing heavy metals by membrane distillation. *J. Hazard. Mater.* 37(1), 77–82. doi:10.1016/0304-3894(94)85035-6.

3

Theoretical Approach behind Membrane Processing Techniques

Komal Parmar

Contents

3.1	Introduction	97
3.2	Microfiltration	100
3.3	Nanofiltration	103
3.4	Ultrafiltration	106
3.5	Reverse Osmosis	107
3.6	Electrodialysis	110
3.7	Membrane Modules	111
References		112

3.1 INTRODUCTION

Membranes can be defined in numerous ways, which can vary broadly in terms of clarity and inclusiveness. In general, a membrane is defined as a layer of semi-permeable material that is used to separate solute material on the application of trans-membrane pressure. Different definitions have been provided in previous writings, a large portion of them concentrating on the particular characteristics of the membrane. Strathmann defined a membrane as an inter-phase that separates two phases and limits the transport of various elements in a particular mode (Strathmann 1986). Drioli and Giorno defined a synthetic membrane as an irregular inter-phase between two phases that allows for the exchange of energy, matter, and information with a selective or non-selective process (Drioli and Giorno 2016). Ulbricht defined a membrane as an inter-phase between two phases in two particular compartments with their permaselective barrier characteristics enabling the exchange of mass and various other applications (Ulbricht 2015).

Membrane technology employs the selective/non-selective characteristics of a membrane, which can thereby be utilized in various applications. Membrane processing technique helps in the extraction, separation, and concentration of solute molecules without the involvement of heat. It is principally associated with reverse osmosis, nanofiltration, microfiltration, ultrafiltration, and electrodialysis for various applications, including the concentration of small and large solutes, ions, removal of bacteria, and others. Further, the advantage of membrane technology is that it works on a minimum requirement of energy without any employment of materials, which can eventually interact with the solute molecules and further change their chemistry. The size of the pores in the membrane is the major factor affecting the transport of solute molecules across the membrane.

Membrane technology has been used in the food industry for various specific applications for more than 50 years. It comes under the category of green technologies. It is expected to observe growth on account of a boost in the application in the food processing, biotechnology, and pharmaceutical industries. In addition, the technology can also be utilized for environmental conservation remedies and energy recovery processes (Le and Nunes 2016; Długołęcki and van der Wal 2013). In the food industry, membrane processing techniques are utilized for a wide range of applications, which include deacidification, demineralization, desalination, separation of solvent, separation of microbial load, and other unwanted solutes. The advantages of membrane technology over traditional methods are overwhelming. For example, sterilization using a suitable membrane is more appropriate for thermolabile products instead of exposure to high temperatures. In addition, it is more economical in terms of energy utilization (Onsekizoglu 2015; Bevilacqua et al. 2018). Sterilization using this method instead of adding preservatives and other additives improves the shelf life of the food product and creates a environmentally safe image. The concentration of a solution using a membrane process retains the natural aroma and nutritional value of the food, which might otherwise be lost on the application of heat (Cros et al. 2005; Paz et al. 2017; Carvalho and Silva 2010). By using membrane technology, wastewater can be processed prior to disposal or for reuse, and this helps to reduce the economic burden of making clean water available every time and further helps in environmental conservation (Peters 2010; Zhang et al. 2016; Bottino et al. 2009). Membrane technology used for commercial purposes is listed in Table 3.1.

Membrane processes are driven by trans-membrane/hydraulic pressure applied on either side of the membrane. When the product is passed through the membrane, one part remains on the membrane, which is called the concentrate, and another part filters out, which is called the filtrate. The product of interest could be in either portion, i.e., concentrate or filtrate or in both. The trans-membrane pressure generated and the pore

Table 3.1 Various Membrane Technologies Used in Various Food Processing Industries

Membrane Technology	Membrane Used	Manufacturer	Application
Nanofiltration	Proprietary PA TFC	Synder® Filtration, USA	Dairy processing, Water treatment
Nanofiltration	HYDRACoRe10, HYDRACoRe50	Hydranautics®, Japan	For colour adjustments and removal
Ultrafiltration	XT	Synder® Filtration, USA	Dairy processing, Water treatment, biotechnology/ pharmaceutical product preparation, food and beverage processing
Ultrafiltration	ZeeWeed 1500	Suez® Water technologies and solutions, USA	Water treatment
Microfiltration	FR	Synder® Filtration, USA	Dairy processing, Water treatment, food and beverage processing
Reverse osmosis	Dow Filmtec BW30-400-34	Dupont® Water solutions, USA	Water treatment
Electrodialysis	ACILYZER EDR	ASTOM® Corporation, Japan	Water treatment

APPLICATIONS OF MEMBRANE TECHNOLOGY

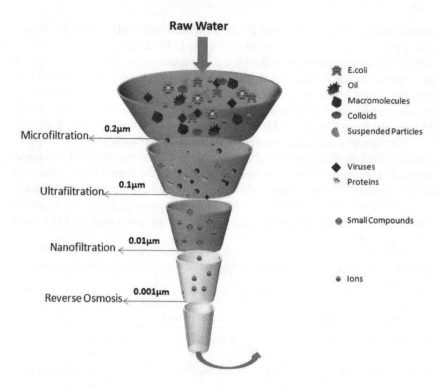

Figure 3.1 Various types of membrane processing techniques.

size of the membrane differentiates the processes like microfiltration, nanofiltration, ultrafiltration, and reverse osmosis (Kotsanopoulos and Arvanitoyannis 2015) (Figure 3.1). However, apart from trans-membrane pressure and the pore size of the membrane, the charge of the molecule and the affinity of the molecule towards the membrane also play a key role in the movement across the membrane (Childress and Elimelech 2000).

3.2 MICROFILTRATION

The first microfiltration membranes were reported by Frick in 1855, who experimented on the fabrication of cellulose nitrate micromembranes by simply dipping the test tube in a collodion solution (Ismail and Goh 2015). Particles greater than 0.1 μm can be separated using microfilter membranes. Hydrodynamic resistance offered by microfilter membranes is less,

THEORETICAL APPROACH BEHIND MEMBRANE PROCESSING TECHNIQUES

hence the application of low hydrostatic pressure (0.1 to 2 bar) will result in high flux. The basic principle behind filtration using microfiltration membranes is to remove 0.1 to 10 μm particles by simple retention on the membrane without application of high vacuum pressure. Microfiltration membranes are a porous matrix that is suitable to separate large particles, algae, or bacteria and allows for the passage of water, monovalent ions, dissolved colloidal materials, and viruses. Thus, the passage of particles through the microfilters depends on the pore size of the membrane and the size of the solute particles itself. The larger the particles, the more retention will be observed on the microfilter membranes through the sieving mechanics, and if the particles are smaller than the pores of the membrane, partial separation can be observed depending on the assembly of the membrane. On this basis, membrane filtration can be utilized as a disinfection barrier to remove larger foreign particles, including pathogenic bacteria. The limitations associated with microfilter membranes include that they are relatively uneconomical due to the high operational costs, the wear and tear of costly membranes, and a reduced inefficiency due to membrane fouling, i.e., due to clogging of the membranes after adsorption onto the surface of the pore and formation of cake on the membrane by the large aggregates. This is also known as dead-end filtration. This occurs when the filtration volume increases and the flux rate decreases due to an increase in membrane fouling (Figure 3.2a). This can be further prevented by using crossflow filtration or tubular filtration approach. In crossflow filtration, a consistent tumultuous stream passes along the membrane surface, thereby preventing the conglomerating of the matter.

The process is called a crossflow or tangential flow filtration because the filtration flow is perpendicular to the feed flow in the tube with the layer of membrane on the surface of the inside wall of the tube (Figure 3.3). In this, the feed flow is at high pressure inside the tube, which further aids in the filtration process. The high flow rate creates the generation of tumultuous stream, thereby preventing the blockage of the membrane, which prevents the rapid drop-off in flux rate and allows a higher volume to be filtered (Figure 3.2b) (Hasan et al. 2013; Herterich et al. 2017). Microfiltration membranes are generally made up of various polymers such as cellulose derivatives (Hu et al. 2019), polypropylene (Pi et al. 2016), polyvinylidene fluoride (Chen et al. 2017), polysulfones (Ohya et al. 2009), polyester (Chollom et al. 2017), and nylon (Huang et al. 2013). The selection of an appropriate membrane for microfiltration contributes to the efficiency of the process, which mainly depends on the type of the fluid, pH of the fluid system, the temperature of the dispersion, nature of the

APPLICATIONS OF MEMBRANE TECHNOLOGY

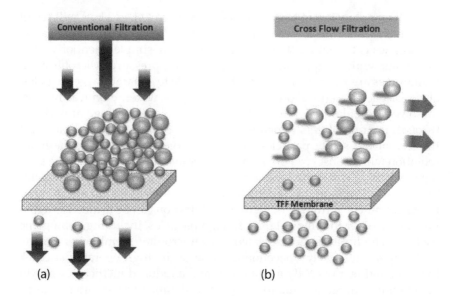

Figure 3.2 Schematic diagram of conventional filtration and crossflow filtration.

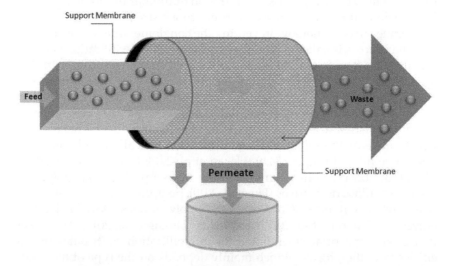

Figure 3.3 Tubular microfiltration membrane module with crossflow filtration.

dissolved solids, molecular weight of the solutes, and loading of the suspended solutes. The selection of the suitable membrane also depends on the cost, percentage recovery, and pretreatment requirements involved for the filtration process.

Microfiltration membrane technology is widely utilized in food processing industry like for water treatment (Koyuncu et al. 2015): Akdemir and Ozers reported the application of a microfiltration process for the treatment of olive oil mill wastewater (Akdemir and Ozers 2006) used for the clarification of fruit juices; Carvalho and Silva reported efficient clarification of pineapple juice using a tubular polyethersulfone membrane at a low transmembrane pressure of 1.5 bar (Carvalho and Silva 2010); Domingues and his co-investigators demonstrated microfiltration of passion fruit juice using hollow fiber membranes of polieterimide (0.4 μm pore size) with a filtration area of 0.056 m² (Domingues et al. 2014) for beer and wine clarification (Oliveira and Barros 2011; De Bruijn 2012; Cimini et al. 2013) and processing of edible oils; Majid and co-researchers demonstrated the processing of crude palm oil using a microfiltration membrane technology (Majid et al. 2000).

3.3 NANOFILTRATION

Nanofiltration membranes have pore sizes in the nanometer range that fall between 1 and 10 nm, and thus they can facilitate the filtration of small ions. Nanofiltration is a pressure-driven process employing an osmosis and ultrafiltration mechanism; with membrane cut off of 200 to 1,000 Dalton (Peeva et al. 2010; Roth et al. 2014). Thus, a nanofiltration membrane removes solute molecules with higher molecular weights. Transportation through membranes can be demonstrated using models. Well-known models include the pore-flow and solution diffusion models, which are associated with a function of the physical and chemical properties of a nanofiltration membrane (Figure 3.4). The pore-flow model assumes that the concentrations of solvent and solute inside a membrane are the same. The chemical potential gradient across the membrane is exhibited as a pressure gradient (Figure 3.5a). The solute diffusion model assumes that pressure everywhere inside the membrane is constant. The chemical potential gradient across the membrane is demonstrated as a concentration gradient (Figure 3.5b). The models can be expressed with the following equation:

$$Ji = -Li \cdot \frac{d\mu i}{dx}$$

APPLICATIONS OF MEMBRANE TECHNOLOGY

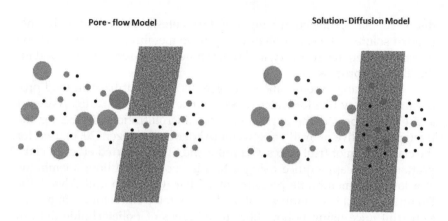

Figure 3.4 Schematic diagram of a pore-flow model and a solution-diffusion model.

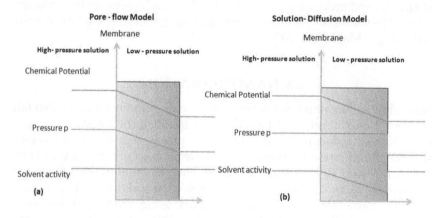

Figure 3.5 Pressure and concentration driven permeation of one-component solution.

Where, Ji is permeate flux of the component i, μi is the total chemical potential of species i and Li is the proportionality coefficient between the flux and driving force.

The Donnan effect is observed in nanofiltration (Hu et al. 2018); the membranes are mostly negatively charged and can filter multivalent ions (about 99%), monovalent ions (about 70%), and organic compounds (about 90%). The mechanism of nanofiltration relies on the membrane structure and on the interlinkage between the membrane surface and the solute molecules to be filtered. The mechanism of ionic transport in nanofiltration was reported by Agboola and co-researchers, which demonstrated various models. The Nernst-Plank equation exhibited an involvement in charge, steric, and dielectric effects in the nanofiltration membrane process (Agboola et al. 2015). The electrostatic interaction of the charged molecules dominates the filtration process when using nano-filter membranes (Figure 3.6). Zou and co-researchers reported the removal of nitrate impurities from groundwater (Zou et al. 2018). However, positively charged nanofiltration membranes are also fabricated to separate positive ionic impurities following a similar mechanism of filtration (Zhang et al. 2017). Fang and co-researchers reported a positively charged nanofiltration membrane fabricated from a polyvinyl chloride graft poly (N, N-dimethylaminoethyl methacrylate). The membrane exhibited excellent water permeability and salt rejection of about 93% (Fang et al. 2019). Li and co-workers fabricated novel positively charged nanofiltration membrane fabricated with the reaction of carboxylic acids on the surface of a polyamide thin film composite with poly(amidoamine) dendrimer in the presence of 2-chloro-1-methylpyridinium iodide as an activating agent. The membrane exhibited excellent rejection efficiency of the positively charged toxic elements (Li et al. 2017). A nanofiltration membrane consists of two layers: one as a supporting barrier membrane and the other as a microporous sublayer. Materials for a nanofiltration membrane includes cellulose acetate (Omidvar et al. 2015), aromatic polyamides (Ahmad et al. 2004), sulfonated polyethersulfone (Ghosh et al. 2002), polypiperazine (Misdan et al. 2015), and mixed matrix membranes (Venkatesh et al. 2016; Gholami and Mahdavi 2018).

Membrane fouling is a critical challenge associated with nanofiltration, which hinders the efficient filtration process. This results due to the

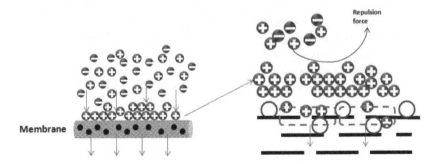

Figure 3.6 Steric effect in nanofiltration membrane technology.

APPLICATIONS OF MEMBRANE TECHNOLOGY

accumulation of unwanted materials on the surface of the membrane, thereby blocking the permeability. Materials might include microbes, inorganic/organic colloids, and precipitated salts. This can further influence the rejection performance of the membrane due to excess polarization within the concentrated material on the surface (Tang et al. 2011).

The nanofiltration membrane has wide applications, including wastewater treatment (Abdel-Fatah 2018), food industry (Nath et al. 2018), and pharmaceutical processes (Chen et al. 2017). Mucchetti and his co-workers reported the application of nanofiltration technology in the production of Quarg-type cheese. In the process, milk pre-concentration was carried out using a nanofiltration membrane, which yielded an acceptable permeation flux of 10 to 41 $kg.h^{-1}.m^{-2}$. The cheese prepared from the processed milk produced a sweeter cheese as compared to the traditional fresh cheese (Mucchetti et al. 2000).

3.4 ULTRAFILTRATION

Ultrafiltration is the process of membrane filtration that acts as a barrier and separates high molecular weight materials, organic and inorganic polymer molecules, suspended solids, colloidal matter, bacteria, viruses, and other unwanted harmful foreign materials from the solvent. The pore size of the membrane used in an ultrafiltration process ranges from 0.1 to 0.001 micron. The membrane is not efficient for removing low molecular weight organics and ion species such as sodium, calcium, magnesium and sulfates. The principle associated with the filtration process by ultrafiltration membrane technology involves low hydrostatic pressure applied across the membrane during filtration. Early ultrafiltration methods involve passing the water sample under pressure through aluminium/lanthanum alginate filters (Poynter et al. 1975). These have the specific advantage of being solubilized in isotonic sodium citrate solution. Advanced techniques involve passing the sample through capillaries, hollow fibers, or membranes with a permissible pore size that allows the water and low molecular weight substances to permeate but reject the macromolecules.

The ultrafiltration process is a crossflow filtration process in which the flow of the sample is tangential to the membrane surface. The main mechanism of filtration is sieving or size exclusion, which removes particles in size ranges of less than 0.01 to 0.1 mm. The performance of the

106

THEORETICAL APPROACH BEHIND MEMBRANE PROCESSING TECHNIQUES

membrane is demonstrated by its retention of large molecules and its filtration rate. The following equation describes the process:

$R = [(Cf–Cp)/Cf \backslash \; x \; 100$

Where, Cf is the concentration of a component of the feed, and Cp is the concentration of that component in the permeate and R is the retention of the solute on the membrane.

Polymers used in the ultrafiltration membrane include polysulfone (Zodrow et al. 2009), polyethersulfone (Li et al. 2015), polyvinylidene difluoride (Hong and He 2014), and cellulose acetate (Saraswathi et al. 2019). Fouling is an intrinsic phenomenon that leads to a reduction in flux and membrane life. Various factors affecting fouling include feed temperature, feed pressure, and feed pH. Mamtani and co-workers reported the effect of various parameters on the fouling of the ultrafiltration membrane. Results suggested that fouling caused by organic substances was higher than caused by inorganic materials. Among the inorganic materials, iron showed more fouling than manganese (Mamtani et al. 2014). The vortex/dean flow filtration is the modification of ultrafiltration technology wherein Taylor vortices are developed in the filtration apparatus by rotation of one cylinder placed into another under pressure (Figure 3.7). This also prevents the surface of the membrane from clogging (Jaffrin 2012). Overall ultrafiltration has the advantage of filtration without any pre-conditions required (Wyn-Jones and Sellwood 2001).

3.5 REVERSE OSMOSIS

Reverse osmosis works on the principle of the obstruction of salt molecules under the influence of externally applied pressure, which is greater than the osmotic pressure. With this mechanism, the membrane will only allow water molecules to permeate from the concentrated solution to the diluted solution (Figure 3.8).

$$J = A(\Delta\rho - \Delta\pi)$$

Where, J is the flux, A is a constant depending on the physical characteristic of the membrane, $\Delta\rho$ is the pressure difference across the membrane, and $\Delta\pi$ is the osmotic pressure difference between the feed and the permeate (Shenvi et al. 2015).

Reverse osmosis membranes have a pore size of less than 1 nm and require very high pressure of up to 80 bar to overcome the osmotic

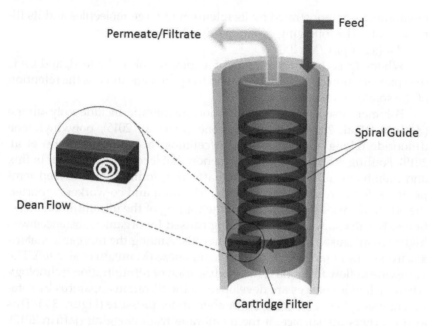

Figure 3.7 Schematic representation of vortex/dean flow in ultrafiltration apparatus.

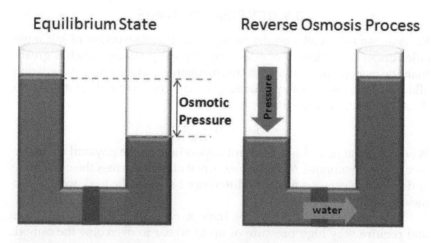

Figure 3.8 Schematic representation of the reverse osmosis process.

THEORETICAL APPROACH BEHIND MEMBRANE PROCESSING TECHNIQUES

pressure of the concentrated solution (Fritzmann et al. 2007). The performance of the reverse osmosis membrane is governed by many factors *viz.* flow rate, permeate flux, salt rejection, recovery rate, differential pressure, and transmembrane pressure. These factors affect the performance of the reverse osmosis membrane filtration process (Table 3.2). Three types of flow rate exist in the process. Feed flow rate is the flow of water entering the reverse osmosis system; permeate flow rate is the flow of the water passing through the membrane, and concentrate flow rate is the flow rate of the concentrate, which is not passed through the membrane. Permeate flux demonstrates the amount of permeate produced during membrane filtration in unit time and unit area. Salt rejection is the amount of solute retained by the membrane during filtration. The recovery rate is a fraction of the feed flow that passes through the membrane. Differential pressure is the pressure drop between the feed and concentrate pressure during water flow across the membrane.

Transmembrane pressure consists of the pressure difference between the feed and the permeate side of the membrane. Reverse osmosis membranes are cellulose-based (Duarte et al. 2006), polyamide (Wei et al.

Table 3.2 Influence of Operating Conditions on Performance of Reverse Osmosis Membrane Process

Operating conditions (If increased)	Influence on	
	Permeation	**Retention of solute**
Feed pressure	Permeation of solvent will increase	Retention of solute will also increase
Feed concentration	Because of the high amount of solute, permeation of solvent will decrease	Whereas retention will increase
Temperature of the system	Permeation of the solvent will increase with a decrease in viscosity of the solvent	Whereas retention will decrease, as an increase in temperature can lead to easy permeation of the solute molecules through the membrane pores
Recovery (relation between permeate flow and feedwater flow)	Permeate flux will decrease	Solute retention will decrease

2016), inorganic/ceramic (Kwak et al. 2001), and mixed matrix membranes (Raza et al. 2019). Reverse osmosis membranes are likely to show fouling in different forms *viz.* biofouling, organic fouling, inorganic scaling, and colloidal fouling (Jiang et al. 2017). From some studies it is found that surface modification, such as improving surface smoothness, hydrophilicity can slow down membrane fouling (Choudhury et al. 2018). Jee and co-workers reported improved the fouling resistance of a polyamide reverse osmosis membrane. Surface modification of the membrane was carried out using a polysiloxane system consisting of 3-glycidoxypropyltrimethoxysilane. Further, a hydrophilic epoxy compound was used, which forms hydrolyzed functional groups and thus reduces fouling (Jee et al. 2016).

3.6 ELECTRODIALYSIS

Electrodialysis is a membrane technology for filtration in which the ions are carried through the ion-selective semi-permeable membrane, under the influence of an electric field (Figure 3.9). It is the combination of the electrolysis and dialysis technique. In this system, the ion-exchange membranes are set up between the anode and cathode. Ions present in

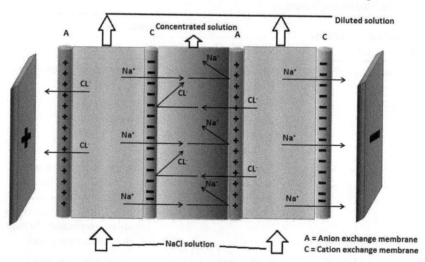

Figure 3.9 Schematic representation of the electrodialysis membrane process.

the feed solution will move toward the ion-exchange membrane under the influence of the electric potential. The cations flow toward the cathode (negative electrode), while the anions move toward the anode (positive electrode). In this process, the cations will pass through the cation exchange membrane but will be collected on the anion exchange membrane.

Similarly, anions will pass through the anion exchange membrane but will be collected on the cation exchange membrane. In this manner, dissolved solutes in ionic forms are selectively separated using ion-exchange membranes. The membrane is made with a combination of ion-exchange resins with polymer-like polypropylene (Ghafoor et al. 2017) or polystyrene (Sachdeva et al. 2008) or by block polymerization of polyelectrolyte (Liu et al. 2010). A demerit of this technique is that it is limited to the compounds which ionize and thus can be blocked by organics, colloids, and biomasses, which will reduce the transfer of ions (Lee et al. 2009). Pretreatment of the feed solution can prevent fouling in an electrodialysis membrane process.

3.7 MEMBRANE MODULES

Membranes are available in different types of modules named plate and frame, tubular, spiral wound, and hollow fiber (Figure 3.10). Membrane modules are the separation units in which the membranes are fitted. The plate and frame is the simplest form of membrane module. It consists of two end plates, the membrane, and the spacers. The membrane and spacers are sandwiched between the two endplates. The feed is forced across the surface of the membrane, which will then pass through the membrane and enter the permeate channel. Tubular modules are made up of several tubes with porous walls in which the membrane is often inside the tube, and the feed is passed through the tube. It works through tangential crossflow. The spiral wound is the advanced version of a plate and frame module. In this module, the membrane sheet is enfolded around the permeate channel. The feed flows through the membrane, and the permeate gets collected in the permeate channel, thereby it moves further in a spiral direction toward the central collection tube. Hollow fiber modules consist of a group of hollow fibers in the pressure vessels. Each pressure vessel consists of many such groups of hollow fibers. This leads to high packing density and even permeate flow with a low-pressure drop.

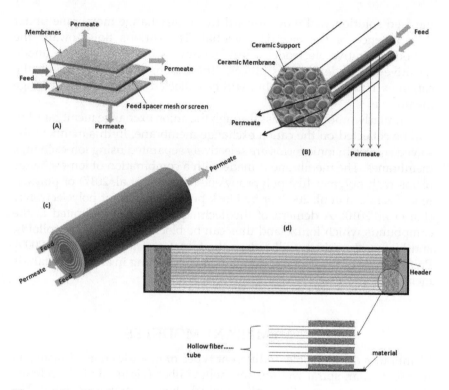

Figure 3.10 Membrane modules (a) Plate and frame (b) Tubular (c) Spiral wound (d) Hollow fiber.

REFERENCES

Abdel-Fatah MA (2018) Nanofiltration systems and applications in wastewater treatment: Review article. *Ain Shams Engineering Journal*, 9(4): 3077–3092.

Agboola O, Maree J, Kolesnikov A, Mbaya R, Sadiku R (2015) Theoretical performance of nanofiltration membranes for wastewater treatment. *Environmental Chemistry Letters*, 13(1): 37–47.

Ahmad AL, Ooi BS, Mohammad AW, Chodhury JP (2004) Composite nanofiltration polyamide membrane: A study on the diamine ratio and its performance evaluation. *Industrial and Engineering Chemistry Research*, 43(25): 8074–8082.

Akdemir EO, Ozers A (2006) Application of microfiltration process to the treatment of olive oil mill wastewater. *Electronic Journal of Environmental, Agricultural and Food Chemistry*, 5(3): 1338–1348.

Bevilacqua A, Petruzzi L, Perricone M, Speranza B, Campaniello D, Sinigaglia M, Corbo MR (2018) Nonthermal technologies for fruit and vegetable juices and beverages: Overview and advances. *Comprehensive Reviews in Food Science and Food Safety*, 17(1): 2–62.

Bottino A, Capannelli G, Comite A, Ferrari F, Firpo R, Venzano S (2009) Membrane technologies for water treatment and agroindustrial sectors. *Comptes Rendus Chimie*, 12(8): 882–888.

Carvalho LMJ, Silva CAB (2010) Clarification of pineapple juice by microfiltration. *Ciencia y Tecnologia de los Alimentos*, 30(3): 828–832.

Chen D, Sirkar KK, Jin C, Singh D, Pfeffer R (2017) Membrane-based technologies in the pharmaceutical industry and continuous production of polymer-coated crystals/particles. *Current Pharmaceutical Design*, 23(2): 242–249.

Chen GE, Sun WG, Kong YF, Wu Q, Sun L, Yu J, Xu ZL (2017) Hydrophilic modification of PVDF microfiltration membrane with poly (ethylene glycol) dimethacrylate through surface polymerization. *Journal of Polymer-Plastics Technology and Engineering*, 57(2): 108–117.

Childress AE, Elimelech M (2000) Relating nanofiltration membrane performance to membrane charge (electrokinetic) characteristics. *Environmental Science and Technology*, 34(17): 3710–3716.

Chollom MN, Pikwa K, Rathilal S, Pillay VL (2017) Fouling mitigation on a woven fibre microfiltration membrane for the treatment of raw water. *South African Journal of Chemical Engineering*, 23: 1–9.

Choudhury RR, Gohil JM, Mohanty S, Nayak SK (2018) Antifouling, fouling release and antimicrobial materials for surface modification of reverse osmosis and nanofiltration membranes. *Journal of Material Chemistry A*, 6(2): 313–333.

Cimini A, Marconi O, Moresi M (2013) Rough beer clarification by crossflow microfiltration in Combination with Enzymatic and/or Centrifugal Pretreatments. *Chemical Engineering Transactions*, 32: 1729–1734.

Cros S, Lignot B, Bourseau P, Jaouen P (2005) Reverse osmosis for the production of aromatic concentrates from mussel cooking juices: A technical assessment. *Desalination*, 180(1–3): 263–269.

De Bruijn J (2012) Wine clarification by microfiltration. In: Drioli E, Giorno L (Eds) *Encyclopedia of Membranes*. Springer, Berlin, Heidelberg. DOI:10.1007/978-3-642-40872-4.

Długołęcki P, van der Wal A (2013) Energy recovery in membrane capacitive deionization. *Environmental Science and Technology*, 47(9): 4904–4910. DOI:10.1021/es3053202.

Domingues RCC, Ramos AA, Cardoso VL, Reis MHM (2014) Microfiltration of passion fruit juice using hollow fibre membranes and evaluation of fouling mechanisms. *Journal of Food Engineering*, 121: 73–79.

Drioli E, Giorno L (2016) Membrane (definition, function and structure). In: Drioli E, Giorno L (Eds) *Encyclopedia of Membranes*. Springer-Verlag, Berlin/Heidelberg, Germany. ISBN: 978-3-662-44324-8. DOI:10.1007/978-3-662-44324-8.

APPLICATIONS OF MEMBRANE TECHNOLOGY

Duarte AP, Cidade MT, Bordado JC (2006) Cellulose acetate reverse osmosis membranes: Optimization of the composition. *Journal of Applied Polymer Science*, 100(5): 4052–4058.

Fang L, Zhou M, Cheng L, Zhu B, Matsuyama H, Zhao S (2019) Positively charged nanofiltration membrane based on cross-linked polyvinyl chloride copolymer. *Journal of Membrane Science*, 572: 28–37.

Fritzmann C, Löwenberg J, Wintgens T, Melin T (2007) State-ofthe-art of reverse osmosis desalination. *Desalination*, 216(1–3): 1–76.

Ghafoor B, Hassan MIU, Yasin T, Shabbir S, Hussain SW (2017) Synthesis of cation exchange membrane from polypropylene fabric using simultaneous radiation grafting. *Journal of Space Technology*, 7(1): 20–25.

Gholami N, Mahdavi H (2018) Nanofiltration composite membranes of polyethersulfone and graphene oxide and sulfonated graphene oxide. *Advances in Polymer Technology*, 37(8): 3529–3541.

Ghosh AK, Ramachandhran V, Hanra MS, Misra BM (2002) Synthesis, characterization, and performance of sulfonated polyethersulfone nanofiltration membranes. *Journal of Macromolecular Science, Part A*, 39(3): 199–216.

Hasan A, Peluso CR, Hull TS, Fieschko J, Chatterjee SG (2013) A surface-renewal model of crossflow microfiltration. *Brazilian Journal of Chemical Engineering*, 30(1): 167–186.

Herterich JG, Xu Q, Field RW, Vella D, Griffiths IM (2017) Optimizing the operation of a direct-flow filtration device. *Journal of Engineering Mathematics*, 104(1): 195–211.

Hong J, He Y (2014) Polyvinylidene fluoride ultrafiltration membrane blended with nano-ZnO particle for photo-catalysis self-cleaning. *Desalination*, 332(1): 67–75.

Hu C, Liu Z, Lu X, Sun J, Liu H, Qu J (2018) Enhancement of the Donnan effect through capacitive ion increase using an electroconductive rGO-CNT nanofiltration membrane. *Journal of Material Chemistry A*, 6(11): 4737–4745.

Hu MX, Niu HM, Chen XL, Zhan HB (2019) Natural cellulose microfiltration membranes for oil/water nanoemulsions separation. *Colloids and Surfaces A: Physicochemical and Engineering Aspects*, 564: 142–151.

Huang L, Bui NN, Meyering MT, Hamlin TJ, McCutcheon JR (2013) Novel hydrophilic nylon 6,6 microfiltration membrane supported thin film composite membranes for engineered osmosis. *Journal of Membrane Science*, 437: 141–149.

Ismail AF, Goh PS (2015) Microfiltration membrane. In: Kobayashi S, Müllen K (Eds) *Encyclopedia of Polymeric Nanomaterials*. Springer, Berlin, Heidelberg. DOI:10.1007/978-3-642-29648-2.

Jaffrin MY (2012) Hydrodynamic techniques to enhance membrane filtration. *Annual Review of Fluid Mechanics*, 44(1): 77–96.

Jee KY, Shin DH, Lee YT (2016) Surface modification of polyamide RO membrane for improved fouling resistance. *Desalination*, 394: 131–137.

Jiang S, Li Y, Ladewig BP (2017) A review of reverse osmosis membrane fouling and control strategies. *The Science of the Total Environment*, 595: 567–583.

Kotsanopoulos KV, Arvanitoyannis IS (2015) Membrane processing technology in the food industry: Food processing, wastewater treatment, and effects on physical, microbiological, organoleptic, and nutritional properties of foods. *Critical Reviews in Food Science and Nutrition*, 55(9): 1147–1175.

Koyuncu I, Sengur R, Turken T, Guclu S, Pasaoglu ME (2015) Advances in water treatment by microfiltration, ultrafiltration, and nanofiltration. In: Basile A, Cassano A, Rastogi NK (Eds) *Advances in Membrane Technology for Water Treatment, Materials, Processes and Applications*. Elsevier, pp. 83–128. ISBN: 978-1-78242-121-4.

Kwak SY, Kim SH, Kim SS (2001) Hybrid organic/inorganic reverse osmosis (RO) membrane for bactericidal anti-fouling. 1. Preparation and characterization of TiO2 nanoparticle self-assembled aromatic polyamide thin-film-composite (TFC) membrane. *Environmental Science and Technology*, 35(11): 2388–2394.

Le NL, Nunes SP (2016) Materials and membrane technologies for water and energy sustainability. *Sustainable Materials and Technology*, 7: 1–28.

Lee HJ, Hong MK, Han SD, Cho SH, Moon SH (2009) Fouling of an anion exchange membrane in the electrodialysis desalination process in the presence of organic foulants. *Desalination*, 238(1–3): 60–69.

Li M, Lv Z, Zheng J, Hu J, Jiang C, Ueda M, Zhang X, Wang L (2017) Positively charged nanofiltration membrane with dendritic surface for toxic element removal. *ACS Sustainable Chemistry and Engineering*, 5(1): 784–792.

Li X, Li J, Bruggen BV, Sun X, Shen J, Han W, Wang L (2015) Fouling behavior of polyethersulfone ultrafiltration membranes functionalized with sol–gel formed ZnO nanoparticles. *RSC Advances*, 5(63): 50711–50719.

Liu G, Dotzauer DM, Bruening M (2010) Ion-exchange membranes prepared using layer-by-layer polyelectrolyte deposition. *Journal of Membrane Science*, 354(1–2): 198–205.

Majid RA, Baharin BS, Ahmadun FR, Man YBC (2000) Processing of crude palm oil with ceramic microfiltration membrane. *Journal of Food Lipids*, 7(2): 113–126.

Mamtani VS, Bhattacharya KP, Prabhakar S, Tewari PK (2014) Fouling studies of capillary ultrafiltration membrane. *Desalination and Water Treatment*, 52(1–3): 542–551.

Misdan N, Lau WJ, Ong CS, Ismail AF, Matsuura T (2015) Study on the thin film composite poly(piperazine-amide) nanofiltration membranes made of different polymeric substrates: Effect of operating conditions. *Korean Journal of Chemical Engineering*, 32(4): 753–760.

Mucchetti G, Zardi G, Orlandini F, Gostoli C (2000) The pre-concentration of milk by nanofiltration in the production of Quarg-type fresh cheeses. *Le Lait, INRA Editions*, 80(1): 43–50. DOI:10.1051/lait:2000106.

Nath K, Dave HK, Patel TM (2018) Revisiting the recent applications of nanofiltration in food processing industries: Progress and prognosis. *Trends in Food Science and Technology*, 73: 12–24.

Ohya H, Shiki S, Kawakami H (2009) Fabrication study of polysulfone hollow-fiber microfiltration membranes: Optimal dope viscosity for nucleation and growth. *Journal of Membrane Science*, 326(2): 293–302.

APPLICATIONS OF MEMBRANE TECHNOLOGY

Oliveira RCD, Barros STDD (2011) Beer Clarification with polysulfone membrane and study on fouling mechanism. *Brazilian Archives of Biology and Technology*, 54(6): 1335–1342.

Omidvar M, Soltanieh M, Mousavi SM, Saljoughi E, Moarefian A, Saffaran H (2015) Preparation of hydrophilic nanofiltration membranes for removal of pharmaceuticals from water. *Journal of Environmental Health Science and Engineering*, 13: 42. DOI:10.1186/s40201-015-0201-3.

Onsekizoglu Bagci P (2015) Potential of membrane distillation for production of high quality fruit juice concentrate. *Critical Reviews in Food Science and Nutrition*, 55(8): 1098–1113.

Paz AI, Blanco CA, Andres-Iglesias C, Palacio L, Pradanos P, Hernandez A (2017) Aroma recovery of beer flavors by pervaporation through polydimethylsiloxane membranes. *Journal of Food Process Engineering*, 40(6): e12556. DOI:10.1111/jfpe.12556.

Peeva LG, Sairam M, Livingston AG (2010) Nanofiltration operations in nonaqueous systems. *Comprehensive Membrane Science and Engineering*, 2: 91–113.

Peters T (2010) Membrane technology for water treatment. *Chemical Engineering and Technology*, 33(8): 1233–1240.

Pi JK, Yang HC, Wan LS, Wu J, Xu ZK (2016) Polypropylene microfiltration membranes modified with TiO2 nanoparticles for surface wettability and antifouling property. *Journal of Membrane Science*, 500: 8–15.

Poynter SFB, Jones HH, Slade JS (1975) Virus concentration by means of soluble ultrafilters. In: Board RG, Lovelock DW (Eds) *Methods for Microbiological Assay*. Academic Press, London, pp. 65–74.

Raza MA, Islam A, Sabir A, Gull N, Ali I, Mehmood R, Bae J, Hassan G, Khan MU (2019) PVA/TEOS crosslinked membranes incorporating zinc oxide nanoparticles and sodium alginate to improve reverse osmosis performance for desalination. *Journal of Applied Polymer Science*, 136(22). DOI:10.1002/app.47559.

Roth CD, Poh SC, Vuong DX (2014) Customization and multistage nanofiltration applications for potable water, treatment, and reuse. In: Street A, Sustich R, Duncan J, Savage N (Eds) *Nanotechnology Applications for Clean Water: Solutions for Improving Water Quality: A Volume in Micro and Nano Technologies*. Elsevier, pp. 201–207.

Sachdeva S, Ram RP, Singh JK, Kumar A (2008) Synthesis of anion exchange polystyrene membranes for the electrolysis of sodium chloride. *American Institute Chemical Engineers Journal*, 54(4): 940–949.

Saraswathi MSSA, Rana D, Alwarappan S, Gowrishankar S, Kanimozhi P, Nagendran A (2019) Cellulose acetate ultrafiltration membranes customized with bio-inspired polydopamine coating and in situ immobilization of silver nanoparticles. *New Journal of Chemistry*, 43(10): 4216–4225.

Shenvi SS, Isloor AM, Ismail AF (2015) A review on RO membrane technology: Developments and challenges. *Desalination*, 368: 10–26.

Strathmann H (1986) Synthetic membranes and their preparation. In: Bungay PM, Lonsdale HK, de Pinho MN (Eds) *Synthetic Membranes: Science, Engineering and Applications*. Springer, Dordrecht, p. 4. ISBN: 978-94-009-4712-2.

Tang CY, Chong TH, Fane AG (2011) Colloidal interactions and fouling of NF and RO membranes: A review. *Advances in Colloid and Interface Science*, 164(1–2): 126–143.

Ulbricht M (2015) Nanoporous polymer filters and membranes, selective filters. In: Kobayashi S, Müllen K (Eds) *Encyclopedia of Polymeric Nanomaterials*. Springer, Berlin, Heidelberg. DOI:10.1007/978-3-642-36199-9_357-1.

Venkatesh K, Arthanareeswaran G, Chandra Bose A (2016) PVDF mixed matrix nano-filtration membranes integrated with 1D-PANI/TiO2 NFs for oil–water emulsion separation. *RSC Advances*, 6(23): 18899–18908.

Wei T, Zhang L, Zhao H, Ma H, Sajib MSJ, Jiang H, Murad S (2016) Aromatic poly-amide Reverse-Osmosis membrane: An optimistic molecular dynamics simulation. *The Journal of Physical Chemistry B*, 120(39): 10311–10318.

Wyn-Jones AP, Sellwood J (2001) A REVIEW: Enteric viruses in the aquatic environment. *Journal of Applied Microbiology*, 91(6): 945–962.

Zhang H, Xu Z, Ding H, Tang Y (2017) Positively charged capillary nanofiltration membrane with high rejection for Mg2 + and Ca2 + and good separation for Mg2 + and Li +. *Desalination*, 420: 158–166.

Zhang L, Zhang P, Wang M, Yang K, Liu J (2016) Research on the experiment of reservoir water treatment applying ultrafiltration membrane technology of different processes. *Journal of Environmental Biology*, 37(5): 1007–1012.

Zodrow K, Brunet L, Mahendra S, Li D, Zhang A, Li Q, Alvarez PJJ (2009) Polysulfone ultrafiltration membranes impregnated with silver nanoparticles show improved biofouling resistance and virus removal. *Water Research*, 43(3): 715–723.

Zou L, Zhang S, Liu J, Cao Y, Qian G, Li Y, Xu Z (2018) Nitrate removal from groundwater using negatively charged nanofiltration membrane. *Environmental Science and Pollution Research*: 1–8. DOI:10.1007/s11356-018-3829-6.

4

Deacidification of Fruit Juices by Electrodialysis Techniques

M. Selvamuthukumaran

Contents

4.1 Introduction   119
4.2 Membranes Used in the Electrodialysis Process   121
     4.2.1 Monopolar Membranes   121
     4.2.2 Bipolar Membrane   122
4.3 Deacidification Process   122
4.4 Effect of Electrodialysis on Physicochemical and Organoleptic
     Properties of Deacidified Juices   125
4.5 Methods for Preventing Fouling of Membranes Used in
     Electro Dialysis Process   126
4.6 Conclusion   127
References   127

4.1 INTRODUCTION

Electrodialysis (ED) is a separation technique where separation occurs through the transportation of charged ions from one solution to another solution while passing through membranes under current (Bazinet et al., 1998). ED is an electrochemical separation technique where the membrane of an electrically charged nature and the electrical potential difference separates the ions from the aqueous solution (Strathmann, 1992;

Bazinet et al., 1998). The separation successfully occurs based on the electrical charges and not on particle size. It has got its wide adaptability in the agro-food industries (Figure 4.1). Cheese whey desalination, cane sugar demineralization, fruit juice deacidification, stabilization of alcoholic beverages, and acid and base production can be successfully performed using this technique. The major component of this technique is the ion-exchange membrane, which is composed of a polymer substance with fixed charges that are covalently bonded with a polymer substance (Strathmann, 1992). There are two different kinds of membrane; first, a membrane possessing fixed negative charges known as a cation exchange membrane, and second a membrane with fixed positive charges known as an anion exchange membrane (Strathmann, 1981). The fruit juices obtained from tropical fruits are a rich source of aroma; their

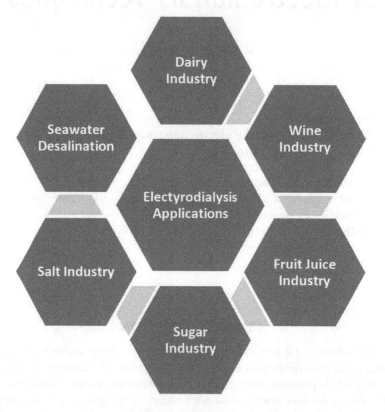

Figure 4.1 Application of the electrodialysis process in various food processing industries.

acidity level should be minimized to enhance their quality when using this technique. The research studies show that this ED technique is successfully employed in the deacidification of fruits juices like pineapple, grapes (Adhikary et al., 1983), orange juice (Goboulev and Salem, 1989), castilla mulberry, etc. (Vera et al., 2007b). The major advantage of this technique is that it does not affect organoleptic attributes (Vera et al., 2007a). The use of bipolar membranes further enhanced stability and performance. The use of mild pressure and temperature conditions retains the valuable constituents as well as the aesthetic appeal of the juice when compared to thermal fruit juice processing.

Adhikary et al. (1983) reported that the acidity of the fruit juices could be minimized using ED stacks with anionic and cationic membranes. They were successfully used for deacidifying fruit juices like orange, pineapple, grape, and lemon. Kang and Rhee (2002) suggested that before carrying out the ED process, initial ultrafiltration of mandarin orange fruit juices significantly minimized acidity content to 30% without affecting sensory as well as chemical constituents like color, vitamin C, flavonoids, etc. The ED stacks can remove citrates such as anions from the juice by passing them through membranes with of an anionic type with two compartments. The citrate ions will be substituted by hydroxyl groups provided by potassium hydroxide solution, which were flowing in the opposite compartments. Voss (1986) reported that an ED stack with a double compartment could efficiently remove acids, as one compartment is made of a bipolar type membrane and the other of an anion exchange type, which can deliver juice with free citric acids.

Calle et al. (2002) and Vera et al. (2003a) found that the pH of passion fruit juices was diminished greatly and was in the range of 2.9–4.0, when compared to the use of chemical agents like calcium carbonate, and they also suggested that the use of bipolar and anionic membranes in the ED process can reduce the citric acid content to a greater extent. The greatest benefit of using ED techniques lies in producing quality enhanced juice; it leads to the usage of chemical reagents, and ultimately production of effluents is highly reduced (Vera et al., 2003b).

4.2 MEMBRANES USED IN THE ELECTRODIALYSIS PROCESS

4.2.1 Monopolar Membranes

There are two different kinds of ion-exchange membranes, which are available: the homogenous membrane and the heterogeneous membrane.

APPLICATIONS OF MEMBRANE TECHNOLOGY

The first is manufactured using phenol polycondensation or phenol-sulphonic acid with formaldehyde. The later is manufactured using dry ion-exchange resins, which are melted and pressed with polymer granules. The membrane consists of greater than 65% w/w of cross-linked ion-exchange particles, which results in poor mechanical strength and dimensional stability; therefore, the use of a heterogeneous membrane is recommended, which is prone to a higher electrical resistance and a higher uneven charge distribution than the homogeneous one.

4.2.2 Bipolar Membrane

There is another well-known and more-frequently used membrane, i.e., the bipolar membrane from the 1980s (Bazinet et al., 1998). Where both the anion and cation exchange membranes were bound together, either physically or chemically, it comprises a thin hydrophilic layer, in which the water molecules can diffuse the aqueous salt solutions from the outside; under an electrical field these molecules are separated into hydroxyl and hydrogen ions (Mani, 1991). There will be no formation of gases either at the surface or within the membranes, which is the major advantage of using a bipolar membrane in the deacidification process. The characteristics of this membrane are that it should possess lower electrical resistance, higher water dissociation rates, ion selectivity, and current density. It should also possess good stability for heat and chemicals when we are using strong acids and bases.

4.3 DEACIDIFICATION PROCESS

The steps involved in the preparation of deacidified fruit juices from tropical fruits are given in Figure 4.2. The deacidification of fruit juice can be performed by carrying out electrodialysis experiments with the help of a laboratory cell with two stack designs, i.e., ED3C (Vera et al., 2007a). This stack can be inbuilt with a homopolar membrane made up of three compartments, in which C1 and C2 are the two compartments in the case shown (Figure 4.3). The stack can also be inbuilt with a homopolar membrane and two bipolar membranes with two compartments, i.e., EDBM2C. The space for each compartment is 0.8 cm, with a similar electrode surface of around 20 cm^2 respective to each membrane area. Selemion CMV (Asahi Glass) is the cation exchange membrane, Neosepta AXE01 (Tokuyama Co.) is the anion exchange membrane and Neosepta

122

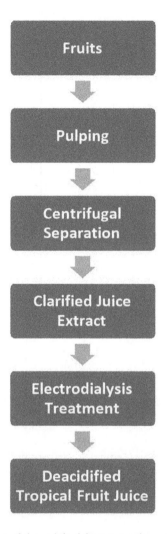

Figure 4.2 Development of deacidified fruit juice from tropical fruits.

BP-1 (Tokuyama Co.) is the bipolar membrane, which can be used for carrying out the deacidification process (Vera et al., 2007a). In an ED3C unit in a juice compartment the clarified juice can be flown with a flow rate of 0.5 dm^3, while in a C1 compartment, 0.1 N Nacl$_2$ can be flown with a flow rate of 2.5 dm^3, in a C2 compartment 0.2N NaOH can be flown with a flow rate 2.5 dm^3, and in the electrode compartment, 0.1 N NaOH can be flown with

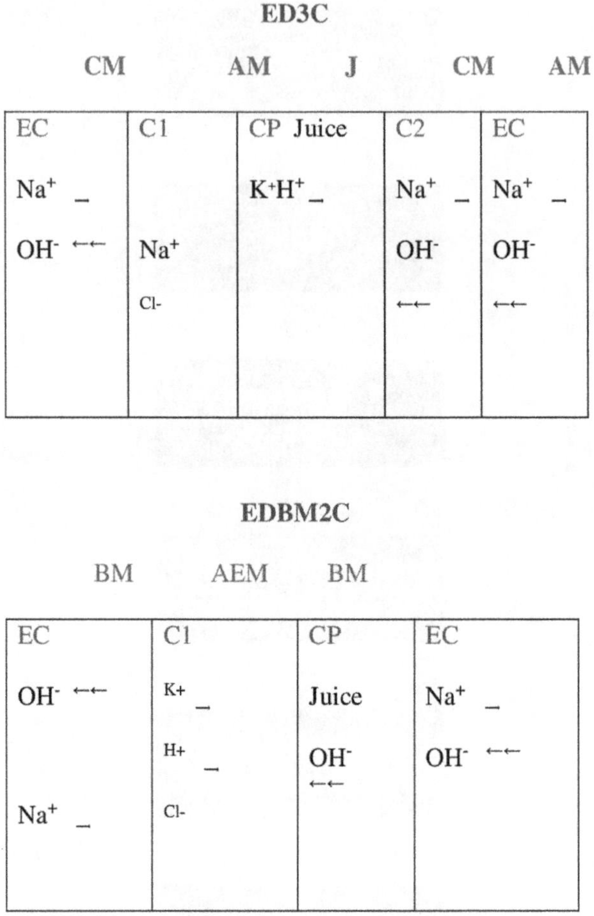
ED3C
CM
AM
J
CM
AM
EC
Na⁺
OH⁻
C1
Na⁺
Cl-
CP Juice
K⁺H⁺
C2
Na⁺
OH⁻
EC
Na⁺
OH⁻
EDBM2C
BM
AEM
BM
EC
OH⁻
Na⁺
C1
K+
H+
Cl-
CP
Juice
OH⁻
EC
Na⁺
OH⁻

Figure 4.3 Electrodialysis cell configurations.

a flow rate of 1 dm^3. The process can be run with a current density of 400 Am^{-2}. The conductivity of voltage and pH were closely monitored during the experiments; the process can be successfully carried out until a pH of 4.0 is obtained during deacidification processes. Two electrodialysis configurations can be used in the citrate anion extraction from juice and their substitution by OH ion groups, which are provided by sodium hydroxide solution in a C2 compartment in ED3C or in EDBM2C, the bipolar membrane, which is sandwiched between juice and electrode compartments.

The first configuration, i.e., ED3C, which consumes more NaOH in the C2 compartment, while the second, i.e., EDBM2C, won't consume the reagent, allowing the cation into the juice to extract and remove only organic and inorganic anions, when juice is flown through the anion exchange membranes. Therefore the juice can be successfully deacidified by configuring electrodialysis membrane processes.

4.4 EFFECT OF ELECTRODIALYSIS ON PHYSICOCHEMICAL AND ORGANOLEPTIC PROPERTIES OF DEACIDIFIED JUICES

Sairi et al. (2004) described that the deacidified pineapple juice using electrodialysis techniques reduced the acidity of pineapple juice from 0.67 to 0.60 for dilute streams and 0.72 for concentrate streams (Table 4.1). The initial process of juice pH was 4.01, and it was slightly minimized to 3.75. It was observed that the buffer salts ionization and elimination during the electrodialysis process play a dominant role in toning or incrementing the pH and thereby reducing the acidity of the pineapple juice. The juice contains several weak acids *viz.* malic acid and citric acid, and salts like calcium, potassium, and sodium. The salts and acids have the capacity to form buffers, and when the juice containing such salts and acids is passed through a membrane the buffering action may be hindered, resulting in a decrease of pH with the reduction of the acid level in the juice (Adhikary et al., 1983). Pineapple juice that passes through the electrodialysis membrane doesn't show remarkable changes in the TSS of the juice.

Table 4.1 Effect of Electrodialytic Treatment on pH and Titratable Acidity of Fruit Juices

Name of the Fruit Juice	Parameter	Before Electrodialysis Treatment	After Electrodialysis Treatment	References
Passion Fruit Juice	pH	2.9	3.3	Vera et al. (2009)
	Titratable acidity	44.6	28.0	
Pineapple Fruit Juice	pH	4.01	3.75	Sairi et al. (2004)
	Titratable acidity	0.67	0.60	

APPLICATIONS OF MEMBRANE TECHNOLOGY

The mineral content of pineapple juice after electrodialysis treatment shows that the zinc content of the juice increased from 0.53 mg% to 0.86 mg%, likewise the iron content from 0.74 to 1.23 mg%, sodium from 0.89 to 2.83 mg%, and aluminum from 0.89 to 1.32 mg%.

The fresh pineapple juice gave a more tart taste when compared to the treated juice; the color and odor didn't bring any remarkable changes, and the electrodialysis treated juice was highly acceptable as plain juice.

4.5 METHODS FOR PREVENTING FOULING OF MEMBRANES USED IN ELECTRO DIALYSIS PROCESS

The presence of both organic and inorganic residues in the membrane system can enhance fouling rate, when membranes were repeatedly used. These residues were left on the internal or external membrane surface, which can reduce the permeation flux of the solute and simultaneously enhance the energy utilization and electrical resistivity nature of the inbuilt ED stacks. To avoid this, the cleaning and maintenance of the membrane should be adequate, which can help with the process of reuse of the membrane at any time. The membrane may lack its free flowingness, and when such organic deposits are identified the molecular movement will be very slow and result in inefficient performance. The foulant solubility is very important, and the utmost care should be taken with the mechanism involved in the precipitation process. The hydrophobic and electrostatic interactions, which were severely affected by the adsorbing materials, the size of the molecules, as well as the pH, have also shown a high effect toward organic acid solubility. If the molecule size is less, solubility and membrane mobility can be enhanced. Lindstrand et al. (2000) explained that there was no correlation coefficient between the total soluble solids of the feed and the cause of fouling; the extent of the fouling can be judged by identifying the type and solubility of the foulant. The different foulants were noted by several authors, i.e., gelatin and starch by De Korosy et al. (1970), grape must by Audinos (1989), and milk whey by Lonergan et al. (1982).

Therefore, in order to prevent the fouling, it is mandatory to pretreat the feed solution either by coagulation (De Korosy et al., 1970) or through integrated membrane techniques like either ultrafiltration or microfiltration (Ferrarini, 2001; Lewandowski et al., 1999; Pinacci et al. 2004), turbulence creation in the compartments, or processing conditions standardization via membrane properties modification (Grebenyuk et al., 1998). Lindstrand et al. (2000) studied the various effects of foulants on charged membranes, and it was shown that they simultaneously enhanced the membrane resistance (Rm) with respect to time.

Membrane fouling can be reduced by using surfactants like oligo-urethanes with different concentrations, and it is also possible to reduce fouling despite the protective layer formation with charges opposite to the membrane base, which does not permit antipolar ions. The use of surfactants for preventing fouling should be harmless, and the leaching of such chemicals should be avoided while processing the food commodities. The H+ and OH– ions should be considered, as they can minimize efficiency. Grebenyuk et al. (1998) reported that enhancing the intensity of alkalinity of the stream can lead to the formation of soluble hydroxides, and such deposits can easily be traced to the cationic membrane during the process of electro-acidification of skimmed milk (Bazinet et al., 2000a). Therefore, one has to take the utmost care that membranes of the anionic type are greatly fouled by organic residues, and membranes of cationic type are fouled by inorganic residues.

4.6 CONCLUSION

The acidity of fruit juices can be toned down to a greater extent by using the electrodialytic process without affecting its organoleptic characteristics. The treated sample showed enhanced mineral content when compared to a thermal method of processing; therefore, it is a novel approach to enhancing the quality of the juice by reducing the acidic content of the fruit juices.

REFERENCES

Adhikary, S.K., Harkare, W.P., and Govindan, K.P. 1983. Deacidification of fruit juices by electrodialysis. *Indian J. Technol.* 21, 120–123.

Audinos, R. 1989. Fouling of ion-selective membranes during electrodialysis of grape must. *J. Membr. Sci.* 41(11), 5–126.

Bazinet, L., Ippersiel, D., Montpetit, D., Mahdavi, B., Amiot, J., and Lamarche, F. 2000. Effect of membrane permselectivity on the fouling of cationic membranes during skim milk electroacidification. *J. Membr. Sci.* 174(1), 97–110.

Bazinet, L., Lamarche, F., and Ippersiel, D. 1998. Bipolar-membrane electrodialysis: Applications of electrodialysis in the food industry. *Trends Food Sci. Technol.* 9(3), 107 –113.

Calle, E.V., Ruales, J., Domier, M., Sandeaux, J., Sandeaux, R., and Pourcelly, G. 2002. Deacidification of the clarified passion fruit juice (*P. edulis f jlavicarpa*). *Desalination* 149(1–3), 357–361.

APPLICATIONS OF MEMBRANE TECHNOLOGY

De Korosy, F., Suszer, A., Komgold, E., Taboch, M.F., Flitman, M., Bandel, E., and Rahav, R. 1970. Membrane fouling and studies on new electrodialysis membranes. *US Office Saline Water, Res. Develop. Progr. Rep. no. 605.*

Ferrarini, R. 2001. A method for tartaric stabilization, in particular for wine, and apparatus for its implementation. *Eur. Pat. Appl.* EP no. 1,146, 115, October, 17th, 2001.

Goboulev, V.N. and Salem, B. 1989. Traitement à l'électrodialyse du jus d'orange. *Ind. Aliment. et Agricoles* 106, 175–177.

Grebenyuk, V.D., Chebotareva, R.D., Peters, S., and Linkov, V. 1998. Surface modification of anionexchange electrodialysis membranes to enhance anti-fouling characteristics. *Desalination* 115(31), 3–329.

Kang, Y.J. and Rhee, K.C. 2002. Deacidification of mandarin orange juice by electrodialysis combined with ultrafiltration. *Nutraceuticals Food* 7(4), 411–416.

Lewandowski, R., Zghal, S., Lameloise, M.L., and Reynes, M. 1999. Purification of date juice for liquid sugar production. *Int. Sugar J.* 101, 125–130.

Lindstrand, V., Sundstrom, G., and Jonsson, A.-S. 2000. Fouling of electrodialysis membranes by organic molecules. *Desalination* 128(1), 91–102.

Lonergan, D.A., Fennemma, O., and Amundson, C.H. 1982. Use of Electrodialysis to improve the protein stability of frozen skim milks and milk concentrates. *J. Food Sci.*, 47 1429–1434.

Mani, K.N. 1991. Electrodialysis water splitting technology. *J M Embr. Sci.* 58, 117–138.

Pinacci, P., Radaelli, M., Bottino, A., and Capannelli, G. 2004. Molasses purification by integrated membrane processes. *Filtration (Coalville, United Kingdom)* 4(2), 119–122.

Sairi, M., Law, Jeng Yih, and Sarmidi, Mohamad Roji 2004. *Chemical Composition and Sensory Analysis of Fresh Pineapple Juice and Deacidified Pineapple Juice Using Electrodialysis.* Universiti Teknologi Malaysia Publisher, 1–8 pp.

Strathmann, H. 1981. Membrane separation processes. *J. Membr. Sci.* 9(1–2), 121–189.

Strathmann, H. 1992. Electrodialysis. In: Winston Ho, W.S. and Sirkar, K.K. eds. *Membrane Handbook.* New York: Van Nostrand Reinhold, pp. 218–262.

Vera, E., Ruales, J., Dornier, M., Sandeaux, J., Sandeaux, R., and Pourcelly, G. 2003a. Deacidifica tion of clarified passion fruit juice using different configurations of electrodialysis. *J. Chem. Techno./. Biotec/mol.* 78, 918–925.

Vera, E., Ruales, J., Dornier, M., Sandeaux, J., Persin, F., Pourcelly, G., et al. 2003b. Comparison of different methods for deacidification of clarified passion fruit juice. *J. Food Eng.* 59(4), 361–367.

Vera, E., Sandeaux, J., Persin, F., Pourcelly, G., Dornier, M., and Ruales, J. 2007a. Deacidification of clarified tropical fruit juices by electrodialysis. Part I. Influence of operating conditions on the process performances. *J. Food Eng.* 78(4), 1427–1438.

Vera, E., Sandeaux, J., Persin, F., Pourcelly, G., Dornier, M., Piombo, G., and Ruales, J. 2007b. Deacidification of clarified tropical fruit juices by electrodialysis. Part II. Characteristics of the deacidified juices. *J. Food Eng.* 78(4), 1439–1445.

Vera, E., Sandeaux, J., Persin, F., Pourcelly, G., Dornier, M., and Ruales, J. 2009. Deacidification of passion fruit juice by electrodialysis with bipolar membrane after different pretreatments. *J. Food Eng.* 90, 67–73.

Voss, H. 1986. Deacidification of citric acid solutions by electrodialysis. *J. Membr. Sci.* 27(2), 165–171.

5

Clarification of Fruit Juices and Wine Using Membrane Processing Techniques

Ismail Tontul

Contents

5.1	Introduction	130
	5.1.1 Turbidity Sources	130
	5.1.2 Steps of Classical Clarification Procedure	133
5.2	Membrane Processing Techniques Used in Clarification of Juices and Wine	133
	5.2.1 Microfiltration (MF)	134
	5.2.2 Ultrafiltration (UF)	134
5.3	Clarification of Juices and Wines Using Membrane Processing	134
	5.3.1 Problems of Membrane Processing	134
	5.3.2 Effect of Membrane Techniques on Permeate Flux	136
	5.3.3 Effect of Membrane Techniques on Quality of Juices and Wines	139
	5.3.3.1 Total or Soluble Solids	139
	5.3.3.2 Microorganisms	140
	5.3.3.3 Vitamins	141
	5.3.3.4 Colour and Pigments	141

5.3.3.5 Phenolics	142
5.3.3.6 Volatile Compounds	143
5.3.4 Effect of Pre-Treatments on Membrane Processing	143
5.4 Novel Approaches in Membrane Clarification of Juices and Wines	145
5.5 Conclusion	147
References	147

5.1 INTRODUCTION

Traditional methods such as clarification and thermal concentration used in fruit juice and wine processing cause significant changes in volatile compounds, vitamins, colour and the nutritive value of products. For this reason, many alternative processing techniques have been tested to prevent or minimise these changes. Figure 5.1 shows traditional and then membrane processing in the processing of fruit juice. Membrane processes are the only promising alternative to the conventional methods for the clarification of fruit juices and beverages. Membrane clarification has many benefits over conventional clarification. For example, the separation of colloid particles does not require a changing temperature and pH or the addition of fining agents. Moreover, membrane clarification reduces the production cost, labour needed, and waste disposal problem (Urosevic et al. 2017). Additionally, membrane clarification allows for the usability of the retained compounds, unlike classical clarification methods (DasGupta and Sarkar 2012).

Extracted or squeezed fruit juices are generally turbid, viscous and dark in colour due to water-insoluble macromolecules and colloid particles (DasGupta and Sarkar 2012). Using mechanical separation systems can only remove coarse particles which are not suitable for sufficient and permanent clarity. The colloid particles have very different physical and chemical properties. Therefore, a combination of physical, biochemical and physicochemical methods must be applied to produce a juice with sufficient and permanent clarity. Clarification of juices and wines is the vital step for an acceptably clear juice or wine. For a successful clarification process, the source of turbidity, clarification conditions and pre-treatments must be known.

5.1.1 Turbidity Sources

Fruit juices contain colloid particles with different chemical properties that cause turbidity problems during processing and storage of the

CLARIFICATION OF FRUIT JUICES AND WINE

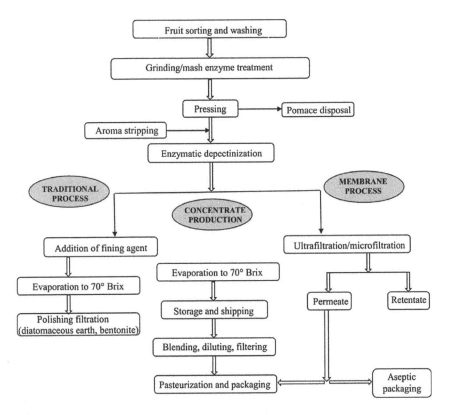

Figure 5.1 Schematic presentation of traditional and membrane processing of juice. (Reprinted from Urosevic et al. (2017) with permission from Elsevier.)

product. Pectin, starch and celluloses are some of the examples of these colloid particles.

Pectic substances are the main colloids in juices. These compounds are found in the cell wall and intercellular layers of the fruits, and composed of galacturonic acid, which may be free or esterified with methanol. Although some fresh fruit juices may contain a higher content of pectin, the average pectin concentration in freshly pressed juices was reported as 0.1–0.2% by Kilara and Van Buren (1989). However, pectic substances act as a protectant of positively charged colloids such as proteins in juices due to the negative charge of the pectins. Therefore, depectinisation using different pectic enzymes is the first step of the classical clarification of fruit juices (Kilara and Van Buren 1989). Depectinisation reduces the viscosity of the juice, has a protective effect and allows the coagulation of suspended

APPLICATIONS OF MEMBRANE TECHNOLOGY

particles. In this way, it increases the efficiency and speed of the filtration of juice (Kilara and Van Buren 1989). There are two types of pectic enzymes, which are depolymerising enzymes and saponifying enzymes. Depolymerising enzymes such as polygalacturonase splits pectic substances and is transformed into a lower molecular weight substance by the cleavage of glycosidic bonds. On the other hand, saponifying enzymes de-esterify the methanol group and yield a low methoxylated pectic substance, which eventually affects the gelation properties (Damodaran and Parkin 2017).

Starches also cause turbidity in fruit juices due to their pasting properties. When starch particles are heated in water, firstly, the particles swell with the binding of water, then leaching of soluble components occurs and, finally, the starch granules become disrupted. This phenomenon is called the pasting of starch (Damodaran and Parkin 2017). The pasted starch dissolves like a colloid in fruit juice and may cross the filter. Therefore, clarified and filtered fruit juice can be cleared even if the starch is not disintegrated. However, these colloidal starch molecules aggregate during storage, and haze-like cloudiness appears in the juice. For these reasons, the starch is hydrolysed enzymatically using different amylases during the depectinisation stage of clarification (DasGupta and Sarkar 2012).

Although fruits only contain a low amount of protein, it can cause serious problems during the filtration and storage of fruit juices. Proteins are positively charged for the pH of most of the fruit juices. As stated above, proteins are coated with pectins in raw juices, and therefore depectinisation is needed to reveal the protein. After depectinisation, proteins are removed using negatively charged fining agents such as bentonite during the clarification process.

Polyphenols are a group of compounds responsible for the taste and colour of fruits and vegetables. These compounds are important for the colour and turbidity of fruit juices and wines. Polyphenols form turbidity in fruit juices by complexing with proteins or the polymerisation and condensation reactions between different polyphenols. Procyanidins are especially highly unstable compounds, and polymerised procyanidins can form high molecular weight molecules to produce turbidity. By removing these high molecular weight compounds, a clear juice with a better taste and improved potability is obtained (Spanos and Wrolstad 1992).

Turbidity caused by the protein–polyphenol interaction is caused by the reaction between proteins capable of binding with proteins (Haze Active Protein) and a polyphenol (Haze Active Polyphenol) capable of bridging the two protein molecules (Siebert et al. 1996). Haze active polyphenols have at least two binding sites and are thus capable of binding to two hydroxy groups in aromatic rings on a polypeptide. Thus, maximum turbidity occurs when the polyphenol binding sites in the structure of the proteins and the binding ends of the polyphenol molecules are equal (Siebert 1999).

5.1.2 Steps of Classical Clarification Procedure

Classical clarification of fruit juices is generally done in two consecutive stages. In the first stage, an enzyme mixture containing mainly pectic enzymes and amylases is added into the juice and left until all of the pectic substances and starches are broken down. This stage of clarification is called depectinisation.

At the second stage, flocculation or clarification of the juice is carried out by adding fining agents. For this purpose, flocculation is provided by adding fining agents to the fruit juice in preliminarily determined amounts. Although there are many methods and agents in this regard, the most commonly used fining agents are gelatin, bentonite and kieselsol.

Gelatin gains a positive charge when added into the fruit juice. Thus, positively charged gelatin flocculates by combining with the negatively charged polyphenols in fruit juice.

The main effect of bentonite on clarification is based on its adsorption power. Although it has positively and negatively charged regions, and the negative load predominates. Therefore, the clarifying effect of bentonite is not only due to adsorption but also because of its charge (Damodaran and Parkin 2017).

Kieselsol is a negatively charged colloidal solution used to remove positively charged colloidal particles from the fruit juices. After the clarification process, fruit juice is filtered using coarse filters.

5.2 MEMBRANE PROCESSING TECHNIQUES USED IN CLARIFICATION OF JUICES AND WINE

Different membrane processes are widely used in the fruit juice and wine industry for various purposes. Nanofiltration is used to purify or enrich compounds that have functional or bioactive properties (Arriola et al. 2014). Reverse osmosis and osmotic distillation are good alternatives for thermal processes to concentrate fruit juices that have a very low impact on the quality and heat-sensitive components (Dincer et al. 2016;

APPLICATIONS OF MEMBRANE TECHNOLOGY

Cassano et al. 2011). Apart from the processes mentioned above, electro-dialysis, pervaporation and diafiltration are used in the food industry. Detailed review papers have been published on the usage of membrane processes in fruit juice (Bhattacharjee et al. 2017; Girard and Fukumoto 2000; Ilame and Singh 2015; Urosevic et al. 2017), wine (El Rayess and Mietton-Peuchot 2016; Mierczynska-Vasilev and Smith 2015) and food industries (Gekas et al. 1998). Microfiltration and ultrafiltration are used for the clarification of fruit juices and wines (Alvarez et al. 1996; Baklouti et al. 2013). Therefore, more detailed information on these techniques will be given in this chapter.

5.2.1 Microfiltration (MF)

MF processes have a pore size of 0.08–2.5 µm and can be used for the separation of particles that have a molecular weight of 150–2500 kDa (Ilame and Singh 2015). Therefore, they are widely used for the clarification of fruit juice, wine and vinegar. MF is a pressure-driven process to retain large colloids, yeast, bacteria, cells and soluble macromolecules. MF membranes are generally made of polymers (polysulfone, cellulose esters etc.) and inorganic materials (ceramic, alumina, silica etc.) (Ilame and Singh 2015)

5.2.2 Ultrafiltration (UF)

UF is widely used to clarify beverages since it can retain particles that have a molecular weight of 80 kDa, such as polymers, colloidal particles, lipids and proteins. Hydrophilic UF membranes generally made from polysulfone, polyethersulfone, polyether ether ketone, etc. The permeate obtained after the UF of fruit juices is composed of amino acids, sugar, salts, phenolics and other water-soluble components (Ilame and Singh 2015). UF has gained popularity in the fruit juice and wine industry because of its advantages, such as improved clarity, with no need to use clarification agents, and it helps in the reduction of microbial loads.

5.3 CLARIFICATION OF JUICES AND WINES USING MEMBRANE PROCESSING

5.3.1 Problems of Membrane Processing

Membrane fouling and concentration polarisation are one of the main problems of membrane processing. Membrane fouling is defined as the alteration in membrane properties because of the interaction of the feed stream and membrane (Mohammad et al. 2012). Four different types of membrane fouling were identified in the literature, namely complete blocking, intermediate blocking, standard blocking and cake filtration (Figure 5.2). When large particles are present in the feed stream generally complete or intermediate blocking occurs. Large particles get trapped on the surface of the membrane and block the pores in both cases. In intermediate blocking other particles were held up by the particles that block the pores (Iritani and Katagiri 2016). In standard blocking, particles smaller than the pores are held up by the internal walls of the membrane and pore volume gradually decreases during filtration (Amosa 2017). Cake formation is the formation of filter cake outside of the membrane surface without blocking the pores (Wang and Tarabara 2008).

Membrane fouling depends on the physicochemical properties of the membrane and feed material to be clarified. Surface morphology of the membrane, solute–solute interaction and feed–membrane interaction affects the nature and extent of membrane fouling (DasGupta and Sarkar 2012). Therefore, optimum clarification conditions (such as membrane material, feed flow rate, transmembrane pressure, etc.) changes according to feed material. Regardless of the fouling mechanism, permeate flux declines throughout the processing, which is the main obstacle in membrane clarification. Therefore, many different studies were conducted to maintain a high flux by changing processing parameters such as transmembrane pressure, feed flow rate, feed temperature and membrane properties.

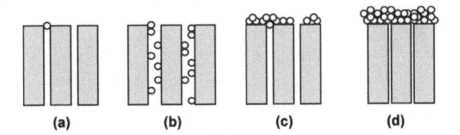

Figure 5.2 Schematics of membrane fouling mechanisms (a) complete blocking, (b) standard blocking, (c) intermediate blocking, and (d) cake filtration (Reprinted from Wang and Tarabara (2008) with permission from Elsevier.)

5.3.2 Effect of Membrane Techniques on Permeate Flux

In many different studies a rapid flux decline at the beginning of membrane clarification was observed due to the deposition and growth of a layer composed of high molecular weight compounds, such as proteins and polysaccharides, on the membrane (He et al. 2007; Li et al. 2006; Mirsaeedghazi and Emam-Djomeh 2017; Mirsaeedghazi et al. 2012; Mondal et al. 2016; Mondor et al. 2000; Nourbakhsh et al. 2015; Rai and De 2009; Severo et al. 2007). A typical flux change during the membrane clarification of lemon juice is given in Figure 5.3.

Transmembrane pressure is one of the most important parameters acting on the permeate flux during membrane clarification. Amirasgari and Mirsaeedghazi (2015) studied the effect of transmembrane pressure (0.1, 0.5 and 1.0 bar) on permeate flux during the clarification of red beet juice using MF. The results of the study showed that increasing the transmembrane pressure resulted in lower irreversible fouling but higher cake resistance. The authors explained this finding with the formation of low-density sediment at the membrane surfaces at low transmembrane

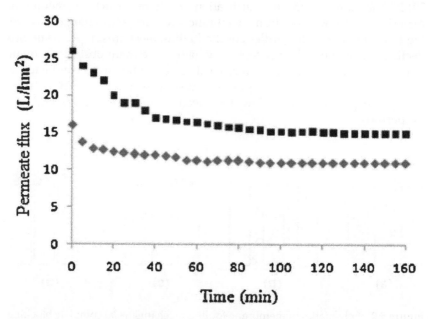

Figure 5.3 Typical permeate flux profile. (Reprinted from Maktouf et al. (2014) with permission from Elsevier.)

pressure. Higher transmembrane pressure increased cake resistance by causing the compression of sediment and the creation of the second layer on the membrane surface (Figure 5.4). Similar results on higher permeate fluxes at high transmembrane pressures were also reported by different researchers (Habibi et al. 2011; Hojjatpanah et al. 2011; Moreno et al. 2012; Nandi et al. 2012; Zarate-Rodriguez et al. 2001). In another study on bottle gourd juice clarification using MF, transmembrane pressure increased the steady-state permeate flux until 60 kPa, and higher pressure than this level did not change the flux (Biswas et al. 2016). This phenomenon was explained using gel layer-controlled filtration. Similar results on gel layer-controlled filtration were also reported for watermelon juice (Chhaya et al. 2008), kiwifruit juice (Cassano et al. 2007), xoconostle juice (Castro-Munoz et al. 2018), black mulberry juice (Hojjatpanah et al. 2011), apple juice (Fukumoto et al. 1998; Li et al. 2006) and banana juice (Sagu et al. 2014).

In a study on red beet juice clarification using microfiltration, increasing the feed flow rate resulted in a higher permeate flux due to an increased tangential force, since it provides a thinner cake at the surface of the membrane (Amirasgari and Mirsaeedghazi 2015). Similar results were also reported for the MF of bottle gourd juice (Biswas et al. 2016), the MF of

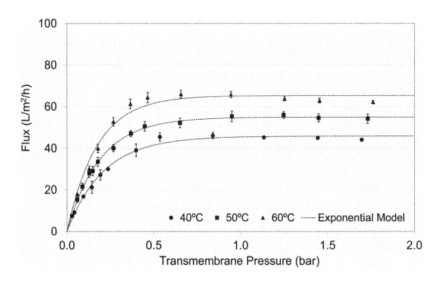

Figure 5.4 Effect of transmembrane pressure on the permeate flux. (Reprinted with permission from Astudillo-Castro 2015. Copyright 2015, American Chemical Society.)

lemon juice (Espamer et al. 2006), the MF or UF of apple juice (Fukumoto et al. 1998), the UF of tomato juice (Razi et al. 2011) and the UF of kiwifruit juice (Cassano et al. 2007; Cassano et al. 2004). Nourbakhsh, Emam-Djomeh, et al. (2014) showed that increasing the feed rate from 0.2 m/s to 0.5 m/s resulted in a 14.3% reduction in cake-formation intensity and an 86.1% reduction in intermediate-blocking intensity. On the other hand, de Bruijn et al. (2003) reported a permeate flux increase effect on the feed flow rate which was dependent upon the applied transmembrane pressure during the UF of apple juice. They determined a slight decrease in the permeate flux when flow rate increased from 2 m/s to 7 m/s at 150 kPa transmembrane pressure. However, an approximately 50% increase in permeate flux was obtained with the increment of feed flow rate of 2 m/s to 7 m/s at 400 kPa of transmembrane pressure.

Feed temperature has a significant effect on permeate flux during membrane clarification (Habibi et al. 2011). A linear increase was determined in the permeate flux during both the MF and UF of apple juice by increasing the feed temperature (Fukumoto et al. 1998). In another study, increasing the feed temperature from 20 to 30°C resulted in a two-fold higher permeate flux during the UF of kiwifruit juice (Cassano et al. 2007). Similarly, permeate flux increased from 15 to 20 $L.m^{-1}.h^{-2}$ by changing the feed temperature from 17 to 29°C during the UF of carrot juice (Cassano et al. 2003). These findings are explained by a reduction of the viscosity of the feed and the increment of the diffusion coefficients of the macromolecules. Nourbakhsh et al. (2014) established a 54% reduction in total resistance with an increase in the feed temperature from 20 to 50°C. According to the results of one study on red plum juice clarification, increasing the feed temperature from 20 to 30°C decreased the intermediate-blocking intensity, while increasing the feed temperature from 30 to 40°C primarily decreased cake-formation intensity (Nourbakhsh, Emam-Djomeh, et al. 2014). On the other hand, Fukumoto et al. (1998) suggested avoiding the use of moderately warm temperatures (20–40°C) due to yeast growth.

Polysulfone (PS) and Polyacrylonitrile (PAN) membranes were compared in the membrane clarification of blood orange juice processing. Results showed that the PS membrane had lower permeate flux than that of the PAN membrane (Conidi et al. 2015). Cassano et al. (2015) compared the effect of hollow fibre polysulfone (PS) and modified poly(ether ketone) (PEEKWC) membranes during the clarification of pomegranate juice. According to fouling analysis and the rejection of phenolics and flavonoids, PSU membranes provided better results than PEEKWC membranes. On the other hand, PEEKWC membranes were found to be more successful in clementine mandarin juice clarification (Cassano et al. 2009). In a study on apple juice clarification, four different polymeric membranes (polyvinylidene fluoride (PVDF), polyethersulfone (PES), PS and cellulose) were compared. According to the flux analysis, PVDF and PS membranes provided a higher flux than other membranes (Girard and Fukumoto 1999). In another study on the UF of raw apple juice using three different membranes (PS 100 kDa, PES 50 kDa and regenerated cellulose 30 kDa), regenerated cellulose membrane was found to be the most efficient material according to fouling characteristics and final product quality (Gulec et al. 2017). Therefore, membrane materials can act differently according to the clarified juice properties.

The hydrophilic and hydrophobic properties of the membrane are other effective factors on the permeate flux during the membrane clarification of juices (DasGupta and Sarkar 2012). Chornomaz et al. (2013) produced poly(vinylidene fluoride) (PVDF) membranes using polyvinylpyrrolidone (PVP) at different ratios to increase the hydrophilic properties of the membrane. Casted membranes were tested in the clarification of lemon juice, and permeate flux was determined. According to the results of the study, casting with PVP provided a 3.5-fold increment in permeate flux compared to the control membrane (without PVP casting).

El Rayess et al. (2012) studied the effect of added tannin, pectin and mannoprotein concentration to filtered wine during MF. Adding tannin, pectin or mannoprotein and increasing the concentration increased permeate flux drastically compared to the filtered wine. Fouling mechanism analysis results showed that tannin addition caused the formation of non-compressible deposit formations on the membrane surface. However, pectin addition caused a compressible deposit. Mannoproteins acted differently from tannins and pectins since the fouling mechanism was identified as complete pore blocking, intermediate pore blocking and standard pore blocking during MF. Among the tested materials, pectins caused the biggest increase on the permeate flux.

5.3.3 Effect of Membrane Techniques on Quality of Juices and Wines

5.3.3.1 Total or Soluble Solids

The microfiltration of red beet juice caused a reduction of total soluble solids from 12° brix to 6.4° brix (Amirasgari and Mirsaeedghazi 2015). In another study on the MF of bottle gourd juice, transmembrane pressure and the feed flow rate were found to be effective on the total solid content of clarified juice (Biswas et al. 2016). The study showed that microfiltration caused a significant reduction in total solids compared to raw juice, and increasing transmembrane pressure and feed flow rate resulted in a decrease in total solids until certain levels.

Different studies have shown that clarification using UF cause a much lower reduction in total or soluble solid content of juices. For example, only a 10% reduction in soluble solid content was reported in xoconostle juice clarification using UF (100 kDa) (Castro-Munoz et al. 2018). Moreover, several studies reported that no significant change in total or soluble solids occur during the UF clarification of different juices (de Bruijn et al. 2003; Zarate-Rodriguez et al. 2001; Vladisavljevic et al. 2013; Ennouri et al. 2015).

Enzymatic pre-treatment with pectinolytic enzymes minimised the rejection of soluble solid content during the UF of mandarin, apple, pear and peach juices (Echavarria et al. 2012).

Chornomaz et al. (2013) cast PVDF membranes with PVP to increase the hydrophilic properties of the membrane and tested the modified membranes in lemon juice clarification. Increasing the hydrophilic properties of the membrane decreased the total soluble solid content of the clarified juices. The UF of ananas juice using PS and ceramic membranes provided a similar reduction on the brix of the product at the end of filtration (de Carvalho et al. 1998).

5.3.3.2 Microorganisms

The microfiltration of roselle extract decreased both total microbial flora count (from 2000 to < 30) and yeasts and moulds count (100 to < 30) (Cisse et al. 2011). In another study, a certified sterilisation MF filter was used to effectively eliminate nine log *Pseudomonas diminuta* (which is a relatively small bacteria) (Fukumoto et al. 1998). The UF of apple juice using a tubular PES membrane with molecular weight cut-off values of 9 or 25 kDa provided a 6–8 log reduction in *P. diminuta* counts (Girard and Fukumoto 1999). Li et al. (2006) successfully sterilised raw apple juice using UF with ceramic membranes under different feed flow rate and transmembrane pressure by providing nine log reductions in *Micrococcus flavus* counts in 1 h (Li et al. 2006). Similarly, four log reductions in the total flora of melon juice was provided using MF. In this way, sterilised melon juice was obtained (Vaillant et al. 2005).

CLARIFICATION OF FRUIT JUICES AND WINE

5.3.3.3 Vitamins

The ascorbic acid content of the microfiltered watermelon juice was not significantly affected by the transmembrane pressure or stirring speed (Chhaya et al. 2008). Similarly, MF of acerola juice (Matta et al. 2004), apple juice (Youn et al. 2004) and melon juice (Vaillant et al. 2005) did not affect ascorbic acid content. In another study on roselle extract, there was a 5% degradation of ascorbic acid content during the MF process. The authors explained this reduction with oxygen exposure of the extract during processing and suggested degassing before microfiltration to prevent ascorbic acid degradation (Cisse et al. 2011). A much higher reduction (22.9%) during the MF of bitter orange juice was reported by Mirsaeedghazi and Emam-Djomeh (2017). Laorko et al. (2010) determined that the ascorbic acid content of the clarified pineapple juice was related to the pore size of the membrane. Smaller pore size resulted in a higher loss of ascorbic acid content of the clarified pineapple juice. Similar results were also reported for the UF of blood orange juice (Toker et al. 2014). Enzymatic pre-treatment of passion fruit juice clarification using MF decreased the ascorbic acid rejection rate (de Oliveira et al. 2012).

5.3.3.4 Colour and Pigments

The microfiltration of red beet juice caused a reduction of betacyanins (5.91%) and betaxanthins (13.04%) (Amirasgari and Mirsaeedghazi 2015). Betalain rejection was reported to be 6% during xoconostle juice clarification using UF (100 kDa) (Castro-Munoz et al. 2018). On the other hand, a 34.5% reduction in betanin content was determined during UF of red beet stalk extract. However, MF caused less rejection (21–22%) of betanin (dos Santos et al. 2016). In another study on beetroot juice clarification using UF, much higher rejection rates for betacyanin and betaxanthin (as 81 and 30%, respectively) were reported (Thakur and Das Gupta 2006).

In different studies, anthocyanin decrement was observed during membrane clarification of juices and wines. For example, an approximately 15% reduction in total anthocyanin content was determined by the MF of raspberry juice (Vladisavljevic et al. 2013). In a study on red plum juice clarification using UF or MF with membranes with different pore sizes, an approximately 30% reduction in anthocyanin content was determined regardless of the pore size or membrane type (mixed cellulose ester or PVDF) (Nourbakhsh et al. 2015). In another study on red plum juice, an approximately 50% reduction was determined in anthocyanin content after UF with PES membrane (100 kDa) (Pap et al. 2012).

141

APPLICATIONS OF MEMBRANE TECHNOLOGY

Conidi et al. (2015) compared total anthocyanin content of blood orange juice clarified with PS or PAN membranes, and a lower total anthocyanin content was determined in the juice clarified with PAN membrane compared to that of PS membrane. A PVDF and mixed cellulose esters membrane caused a similar rejection of anthocyanins (cyanidin-3-glucoside, cyanidin-3,5-diglucoside, delphinidin-3-glucoside, pelargonidin-3-glucoside and pelargonidin-3,5-diglucoside) during the UF of pomegranate juice (Mirsaeedghazi et al. 2010). A similar reduction ratio of total monomeric anthocyanin content was determined by UF or MF of pomegranate juice (Mirsaeedghazi et al. 2012). Prodanov et al. (2013) studied the effect of membrane clarification on the individual anthocyanins of grape pomace extract, and the clarification process was only found to be effective on high molecular weight acylated anthocyanins. Three different membrane materials (PES, cellulose acetate and nylon) with the same pore sizes were compared in the membrane filtration of wine, and a membrane produced from PES had the smallest content of anthocyanins among the tested materials (Urkiaga et al. 2002).

Chhaya et al. (2008) studied the effect of transmembrane pressure and stirring speed on lycopene content during the MF of watermelon juice. Lycopene rejection at the end of the process was found to be in the range of 66.7 and 54.4%. Transmembrane pressure had a significant effect on the lycopene content of clarified watermelon juice, and higher transmembrane pressure resulted in a higher content of lycopene. In another study on melon juice clarification using MF, caused a total rejection of β-carotene in the permeate (Vaillant et al. 2005).

5.3.3.5 Phenolics
Microfiltration of red beet juice (Amirasgari and Mirsaeedghazi 2015), jamun juice (Ghosh et al. 2018) and bottle gourd juice (Biswas et al. 2016) caused an approximately 50% reduction in total phenolic compounds. Much lower rejection of phenolics during the UF of different juices was determined. Castro-Munoz et al. (2018) reported only a 2.10% reduction of xoconostle juice by UF (100 kDa). In a study comparing the MF and UF of pomegranate juice clarification, a reduction of total phenolic content in UF was determined to be lower than MF samples (Mirsaeedghazi et al. 2012). Onsekizoglu et al. (2010) determined an approximately 30% reduction in total phenolic content in apple juice by UF using a 10 kDa membrane. However, the reduction was only 3% when a 100 kDa membrane was used.

Transmembrane pressure was found to be effective on the total phenolic content of UF clarified juices. In a previous study on pomegranate juice clarification with UF (15 kDa), increasing the transmembrane pressure from 1.0 bar to 1.70 bar decreased the total phenolic rejection from 45.45 to 25.72%. On the other hand, higher transmembrane pressure than 1.70 did not significantly affect the total phenolic rejection (Baklouti et al. 2012). Similarly, increasing transmembrane pressure had no effect on total polyphenol retention during the UF of banana juice (Sagu et al. 2014)

Conidi et al. (2015) compared total phenolic content and antioxidant activity of blood orange juice clarified with PS or PAN membranes, and a similar rejection ratio (23%) was observed for both membrane materials.

Gokmen et al. (1998) compared three custom made membranes (using different ratios of PES, PVP, water and 1-methyl-2-pyrrolidone) with two commercial membranes during the UF of apple juice and analysed the total content of polyphenols after clarification. The rejection of the polyphenols was much lower in commercial membranes when compared to custom made membranes. The membrane composition was found to have a significant effect on the polyphenol content of the clarified apple juice. In another study on pomegranate juice, a mixed cellulose esters membrane caused a lower reduction of total phenolic content when compared to PVDF membranes (Mirsaeedghazi et al. 2010). Mondal et al. (2016) produced five different custom membranes (using different ratios of PAN, cellulose acetate phthalate, polyethylene glycol N-dimethylformamide) and compared the effectiveness of the membrane in bottle gourd juice clarification. Their results showed that total phenolic compound retention was related to the composition of the membrane, probably due to the changing hydrophilicity and molecular weight cut-off values of the membrane.

5.3.3.6 Volatile Compounds

Saura et al. (2012) analysed the volatile composition of lemon juice clarified using MF and UF. They used three different pore sizes in membrane clarification. They reported that increasing the filtration time or decreasing the pore size resulted in a higher rejection of volatile compounds in the permeate. The rejection rate of terpene hydrocarbons was the highest, and those were followed by alcohols, aldehydes and esters, respectively.

5.3.4 Effect of Pre-Treatments on Membrane Processing

Bagci (2014) compared the effectiveness of pre-clarification with four different fining agents [gelatin, gelatin+bentonite, polyvinyl polypyrrolidone (PVPP) and PVPP+bentonite] before ultrafiltration (30 kDa) in terms of the permeate flux and physicochemical properties of pomegranate juice.

143

APPLICATIONS OF MEMBRANE TECHNOLOGY

According to the results of the study, pre-clarification with PVPP+bentonite or gelatin+bentonite provided a higher flux during ultrafiltration. Resistance analysis clearly showed that these fining agents significantly reduced the cake resistance by eliminating large particles. The higher flux obtained by these fining agents was related to the reduction of phenolic compounds (ellagic acid, gallic acid, catechin, chlorogenic acid, caffeic acid and ferulic acid) and total monomeric anthocyanins as physicochemical analysis proved (Bagci 2014). On the other hand, pre-clarification did not affect the pH, brix or acid profile of pomegranate juice.

Similarly, Gokmen and Cetinkaya (2007) reported that gelatin and bentonite application before the UF of apple juice decreased the time required for collecting a specific volume of clarified juice. In another study on wine clarification, MF with different treatments (bentonite, activated carbon, zirconium oxide and bentonite + activated carbon) were compared in terms of permeate flux and total polyphenol retention. The highest permeate flux was determined in the MF and bentonite + activated carbon application followed by MF and bentonite application. An approximate 10% reduction of total polyphenols was observed in different treatments (Salazar et al. 2007).

Laccase is an enzyme that catalyses the oxidation of phenolic compounds and causes the formation of melanoidins, which are high molecular weight compounds. Therefore, the effect of laccase treatment on permeate flux during the UF of pomegranate juice was studied by Baklouti et al. (2012). As expected, pre-treatment with laccase increased the permeate flux by approximately 40%. Similar results for enzyme pre-treatment increasing the permeate flux were also reported for the MF of umbu juice (Ushikubo et al. 2006), the UF of pomegranate juice (Neifar et al. 2011) and the UF of apple juice (He et al. 2007). Since laccase catalyses the polymerisation of phenolic compounds, UF or MF after laccase treatment causes a 50% decrease in clarified juice compared to raw juice (Neifar et al. 2011).

Pectinolytic enzyme treatment before the UF of different fruit juices (mandarin, apple, pear and peach juices) provided a 40% increment of the permeate flux (Echavarria et al. 2012). Similar findings were also reported for lemon juice (Maktouf et al. 2014), mosambi juice (Nandi et al. 2009) and raspberry juice (Molnar et al. 2012). Rai et al. (2007) tested the combined usage of bentonite and pectinases on the UF of mosambi juice. The obtained results showed that the combined usage of bentonite and pectinases provided a higher flux compared to the enzyme-treated only samples since total resistance was found to be lower in the enzyme and bentonite applied sample.

In a study conducted to compare the effectiveness of a classical fining agent application with an enzyme treatment to increase permeate flux during membrane clarification of mosambi juice, enzyme application gave a better permeate flux than those of the fining agents (Rai et al. 2006).

5.4 NOVEL APPROACHES IN MEMBRANE CLARIFICATION OF JUICES AND WINES

Sarkar et al. (2008) developed an electric field-assisted UF (Figure 5.5) unit and tested the effect of the electric field, pulse ratio, crossflow velocity and transmembrane pressure during the UF of mosambi juice. Increasing the electric field increased the permeate flux linearly during clarification. A continuation study showed that electric field application reduced energy consumption by 22% (Sarkar et al. 2009). Pulsed application of an electric field was found to be more advantageous when compared to a constant application (Sarkar et al. 2009). Rai et al. (2010) also reported higher flux when the electric field was applied during the UF process.

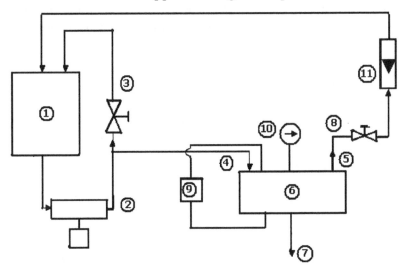

Figure 5.5 Schematic diagram of the electro-ultrafiltration system [1, Feed tank; 2, feed pump; 3, bypass control valve; 4, feed inlet; 5, retentate; 6, cross-flow electro-ultrafiltration; 7, permeate; 8, flow control valve; 9, regulated DC power supply; 10, pressure gauge; 11, rotameter]. (Reprinted from Sarkar et al. (2008) with permission from Elsevier.)

Aghdam et al. (2015) developed a hybrid ultrasonic-membrane system (Figure 5.6) to increase the efficiency of clarification. The system allowed for the reduction of irreversible fouling and cake resistance, which increased the permeate flux. The authors also evaluated the effect of the hybrid system on the physicochemical properties of pomegranate juice. The results showed that total soluble solids, total anthocyanin content, acidity and antioxidant activity were lower in the samples clarified with the hybrid system when compared to membrane clarification without ultrasound. In another study published by the same group, the mechanism of the fouling with or without ultrasound was studied in detail (Aghdam et al. 2015). The study revealed that cake formation was the main fouling mechanism during membrane clarification. However, the ultrasound application caused a change in the blocking index value. According to the results of the study, ultrasonic waves provided a much lower cake formation during membrane clarification.

Youravong et al. (2010) tested gas sparging to reduce concentration polarisation and membrane fouling during the MF of wine. When the gas injection factor (r) was 0.15, the steady flux was 1.4-fold higher than the control treatment (without gas sparging). The reason for this finding was explained by the promoted local mixing due to bubble induced flows.

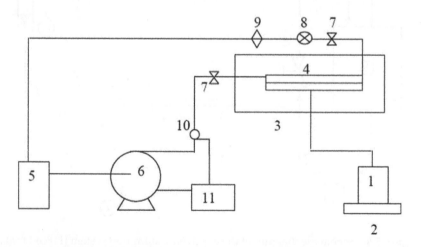

Figure 5.6 Hybrid ultrasonic-membrane unit (1: permeate tank; 2: balance; 3: ultrasonic bath; 4: membrane module; 5: feed tank; 6: pump; 7: pressure meter; 8: valve; 9: flow meter; 10: transmitter; 11: inverter). (Reprinted from Aghdam et al. (2015) with permission from Elsevier.)

Resistance analysis showed a linear decrease of reversible fouling with the increase of the gas injection factor. Irreversible fouling was similar to the control treatment when the gas injection factor was 0.15, but resistance was determined to be higher than the control treatment at higher gas injection factor values. Gas sparging caused a loss of alcohol content from the clarified wines (Youravong et al. 2010).

5.5 CONCLUSION

Membrane processes are becoming increasingly important in the fruit juice industry due to their benefits over conventional techniques. Among different membrane processes, ultrafiltration and microfiltration are widely used for the clarification of fruit juices and wines. According to the obtained results, clarification by ultrafiltration and microfiltration has been demonstrated as a good alternative for the clarification of juices and wines according to product quality, easiness of process and scale-up. Literature studies have shown that UF has less effect on total or soluble solid content, vitamins and phenolics. Additionally, it causes a significant reduction in the microbial count of the clarified product. However, the membrane fouling problem should be prevented, and the durability of the membranes should be increased for the membrane processes to become widespread in the fruit juice and wine industry. Recently, electric field-assisted membrane processing, ultrasound-assisted membrane processing and gas sparging during membrane processing have been developed and reported to improve membrane processes. Development of membranes with less fouling and longer durability, optimisation of the process design and increment of experience will improve the use of membrane technologies for clarification of fruit juices and wines.

REFERENCES

Aghdam, M. A., H. Mirsaeedghazi, M. Aboonajmi, and M. H. Kianmehr. 2015. Effect of ultrasound on different mechanisms of fouling during membrane clarification of pomegranate juice. *Innovative Food Science and Emerging Technologies* 30:127–131.

Aghdam, M. A., H. Mirsaeedghazi, M. Aboonajmi, and M. H. Kianmehr. 2015. The effect of ultrasound waves on the efficiency of membrane clarification of pomegranate juice. *International Journal of Food Science and Technology* 50(4):892–898.

APPLICATIONS OF MEMBRANE TECHNOLOGY

Alvarez, V., L. J. Andres, F. A. Riera, and R. Alvarez. 1996. Microfiltration of apple juice using inorganic membranes: Process optimization and juice stability. *The Canadian Journal of Chemical Engineering* 74(1):156–162.

Amirasgari, N., and H. Mirsaeedghazi. 2015. Microfiltration of red beet juice using mixed cellulose ester membrane. *Journal of Food Processing and Preservation* 39(6):614–623.

Amosa, Mutiu Kolade. 2017. Towards sustainable membrane filtration of palm oil mill effluent: Analysis of fouling phenomena from a hybrid PAC-UF process. *Applied Water Science* 7(6):3365–3375.

Arriola, Nathalia Aceval, Gielen Delfino dos Santos, Elane Schwinden Prudêncio, Luciano Vitali, José Carlos Cunha Petrus, and Renata D. M. Castanho Amboni. 2014. Potential of nanofiltration for the concentration of bioactive compounds from watermelon juice. *International Journal of Food Science and Technology* 49(9):2052–2060.

Astudillo-Castro, C. L. 2015. Limiting flux and critical transmembrane pressure determination using an exponential model: the effect of concentration factor, temperature, and cross-flow velocity during casein micelle concentration by microfiltration. *Industrial & Engineering Chemistry Research* 54(1):414–425.

Bagci, P. O. 2014. Effective clarification of pomegranate juice: A comparative study of pretreatment methods and their influence on ultrafiltration flux. *Journal of Food Engineering* 141:58–64.

Baklouti, S., R. Ellouze-Ghorbel, A. Mokni, and S. Chaabouni. 2012. Clarification of pomegranate juice by ultrafiltration: Study of juice quality and of the fouling mechanism. *Fruits* 67(3):215–225.

Baklouti, S., A. Kamoun, R. Ellouze-Ghorbel, and S. Chaabouni. 2013. Optimising operating conditions in ultrafiltration fouling of pomegranate juice by response surface methodology. *International Journal of Food Science and Technology* 48(7):1519–1525.

Bhattacharjee, C., V. K. Saxena, and S. Dutta. 2017. Fruit juice processing using membrane technology: A review. *Innovative Food Science and Emerging Technologies* 43:136–153.

Biswas, P. P., M. Mondal, and S. De. 2016. Comparison between centrifugation and microfiltration as primary clarification of bottle gourd (*Lagenaria siceraria*) juice. *Journal of Food Processing and Preservation* 40(2):226–238.

Cassano, A., C. Conidi, and E. Drioli. 2011. Clarification and concentration of pomegranate juice (*Punica granatum* L.) using membrane processes. *Journal of Food Engineering* 107(3–4):366–373.

Cassano, A., C. Conidi, and F. Tasselli. 2015. Clarification of pomegranate juice (*Punica granatum* L.) by hollow fibre membranes: Analyses of membrane fouling and performance. *Journal of Chemical Technology and Biotechnology* 90(5):859–866.

Cassano, A., L. Donato, and E. Drioli. 2007. Ultrafiltration of kiwifruit juice: Operating parameters, juice quality and membrane fouling. *Journal of Food Engineering* 79(2):613–621.

Cassano, A., E. Drioli, G. Galaverna, R. Marchelli, G. Di Silvestro, and P. Cagnasso. 2003. Clarification and concentration of citrus and carrot juices by integrated membrane processes. *Journal of Food Engineering* 57(2):153–163.

Cassano, A., B. Jiao, and E. Drioli. 2004. Production of concentrated kiwifruit juice by integrated membrane process. *Food Research International* 37(2):139–148.

Cassano, A., F. Tasselli, C. Conidi, and E. Drioli. 2009. Ultrafiltration of Clementine mandarin juice by hollow fibre membranes. *Desalination* 241(1–3):302–308.

Castro-Munoz, R., V. Fila, B. E. Barragan-Huerta, J. Yanez-Fernandez, J. A. Pina-Rosas, and J. Arboleda-Mejia. 2018. Processing of Xoconostle fruit (*Opuntia joconostle*) juice for improving its commercialization using membrane filtration. *Journal of Food Processing and Preservation* 42(1).

Chhaya, C., P. Rai, G. C. Majumdar, S. Dasgupta, and S. De. 2008. Clarification of watermelon (*Citrullus lanatus*) juice by microfiltration. *Journal of Food Process Engineering* 31(6):768–782.

Chornomaz, P. M., C. Pagliero, J. Marchese, and N. A. Ochoa. 2013. Impact of structural and textural membrane properties on lemon juice clarification. *Food and Bioproducts Processing* 91(C2):67–73.

Cisse, M., F. Vaillant, D. Soro, M. Reynes, and M. Dornier. 2011. Crossflow microfiltration for the cold stabilization of roselle (*Hibiscus sabdariffa* L.) extract. *Journal of Food Engineering* 106(1):20–27.

Conidi, C., F. Destani, and A. Cassano. 2015. Performance of hollow fiber ultrafiltration membranes in the clarification of blood orange juice. *Beverages* 1(4):341–353.

Damodaran, Srinivasan, and Kirk L. Parkin. 2017. *Fennema's Food Chemistry*. CRC Press: Boca Raton, FL.

DasGupta, Sunando, and Biswajit Sarkar. 2012. Membrane applications in fruit processing technologies. In: *Advances in Fruit Processing Technologies*, edited by S. Rodrigues, and F. A. N. Fernandes. Boca Raton: CRC Press.

de Bruijn, J. P. F., A. Venegas, J. A. Martinez, and R. Borquez. 2003. Ultrafiltration performance of Carbosep membranes for the clarification of apple juice. *Lebensmittel-Wissenschaft und--Technologie-Food Science and Technology* 36(4):397–406.

de Carvalho, L. M. J., C. A. B. da Silva, and Aptr Pierucci. 1998. Clarification of pineapple juice (*Ananas comosus* L. Merryl) by ultrafiltration and microfiltration: Physicochemical evaluation of clarified juices, soft drink formulation, and sensorial evaluation. *Journal of Agricultural and Food Chemistry* 46(6):2185–2189.

de Oliveira, R. C., R. C. Doce, and S. T. D. de Barros. 2012. Clarification of passion fruit juice by microfiltration: Analyses of operating parameters, study of membrane fouling and juice quality. *Journal of Food Engineering* 111(2):432–439.

Dincer, Cuneyt, Ismail Tontul, and Ayhan Topuz. 2016. A comparative study of black mulberry juice concentrates by thermal evaporation and osmotic distillation as influenced by storage. *Innovative Food Science and Emerging Technologies* 38:57–64.

APPLICATIONS OF MEMBRANE TECHNOLOGY

dos Santos, C. D., R. K. Scherer, A. S. Cassini, L. D. F. Marczak, and I. C. Tessaro. 2016. Clarification of red beet stalks extract by microfiltration combined with ultrafiltration. *Journal of Food Engineering* 185:35–41.

Echavarria, A. P., V. Falguera, C. Torras, C. Berdun, J. Pagan, and A. Ibarz. 2012. Ultrafiltration and reverse osmosis for clarification and concentration of fruit juices at pilot plant scale. *LWT-Food Science and Technology* 46(1):189–195.

El Rayess, Y., C. Albasi, P. Bacchin, P. Taillandier, M. Mietton-Peuchot, and A. Devatine. 2012. Analysis of membrane fouling during crossflow microfiltration of wine. *Innovative Food Science and Emerging Technologies* 16:398–408.

El Rayess, Y., and M. Mietton-Peuchot. 2016. Membrane technologies in wine industry: An overview. *Critical Reviews in Food Science and Nutrition* 56(12):2005–2020.

Ennouri, M., I. Ben Hassan, H. Ben Hassen, C. Lafforgue, P. Schmitz, and A. Ayadi. 2015. Clarification of purple carrot juice: Analysis of the fouling mechanisms and evaluation of the juice quality. *Journal of Food Science and Technology-Mysore* 52(5):2806–2814.

Espamer, L., C. Pagliero, A. Ochoa, and J. Marchese. 2006. Clarification of lemon juice using membrane process. *Desalination* 200(1–3):565–567.

Fukumoto, L. R., P. Delaquis, and B. Girard. 1998. Microfiltration and ultrafiltration ceramic membranes for apple juice clarification. *Journal of Food Science* 63(5):845–850.

Gekas, V., G. Baralla, and V. Flores. 1998. Applications of membrane technology in the food industry. *Food Science and Technology International* 4(5):311–328.

Ghosh, P., R. C. Pradhan, and S. Mishra. 2018. Clarification of jamun juice by centrifugation and microfiltration: Analysis of quality parameters, operating conditions, and resistance. *Journal of Food Process Engineering* 41(1).

Girard, B., and L. R. Fukumoto. 1999. Apple juice clarification using microfiltration and ultrafiltration polymeric membranes. *Food Science and Technology-Lebensmittel-Wissenschaft and Technologie* 32(5):290–298.

Girard, B., and L. R. Fukumoto. 2000. Membrane processing of fruit juices and beverages: A review. *Critical Reviews in Food Science and Nutrition* 40(2):91–157.

Gokmen, V., Z. Borneman, and H. H. Nijhuis. 1998. Improved ultrafiltration for color reduction and stabilization of apple juice. *Journal of Food Science* 63(3):504–507.

Gokmen, V., and O. Cetinkaya. 2007. Effect of pretreatment with gelatin and bentonite on permeate flux and fouling layer resistance during apple juice ultrafiltration. *Journal of Food Engineering* 80(1):300–305.

Gulec, H. A., P. O. Bagci, and U. Bagci. 2017. Clarification of apple juice using polymeric ultrafiltration membranes: A comparative evaluation of membrane fouling and juice quality. *Food and Bioprocess Technology* 10(5):875–885.

Habibi, A., A. Aroujalian, A. Raisi, and F. Zokaee. 2011. Influence of operating parameters on clarification of carrot juice by microfiltration process. *Journal of Food Process Engineering* 34(3):860–877.

150

He, Yasan, Zhijuan Ji, and Shunxin Li. 2007. Effective clarification of apple juice using membrane filtration without enzyme and pasteurization pretreatment. *Separation and Purification Technology* 57(2):366–373.

Hojjatpanah, G., Z. Emam-Djomeh, A. K. Ashtari, H. Mirsaeedghazi, and M. Omid. 2011. Evaluation of the fouling phenomenon in the membrane clarification of black mulberry juice. *International Journal of Food Science and Technology* 46(7):1538–1544.

Ilame, S. A., and S. V. Singh. 2015. Application of membrane separation in fruit and vegetable juice processing: A review. *Critical Reviews in Food Science and Nutrition* 55(7):964–987.

Iritani, Eiji, and Nobuyuki Katagiri. 2016. Developments of blocking filtration model in membrane filtration. *KONA Powder and Particle Journal* 33:179–202.

Kilara, Arun, and Jerome P. Van Buren. 1989. Clarification of apple juice. In: *Processed Apple Products* (Downing, D.L., editor). Springer.

Laorko, Aporn, Zhenyu Li, Sasitorn Tongchitpakdee, Suphitchaya Chantachum, and Wirote Youravong. 2010. Effect of membrane property and operating conditions on phytochemical properties and permeate flux during clarification of pineapple juice. *Journal of Food Engineering* 100(3):514–521.

Li, J., Z. F. Wang, Y. Q. Ge, Q. F. Sun, and X. S. Hu. 2006. Clarification and sterilization of raw depectinized apple juice by ceramic ultrafiltration membranes. *Journal of the Science of Food and Agriculture* 86(1):148–155.

Maktouf, S., M. Neifar, S. J. Drira, S. Baklouti, M. Fendri, and S. E. Chaabouni. 2014. Lemon juice clarification using fungal pectinolytic enzymes coupled to membrane ultrafiltration. *Food and Bioproducts Processing* 92(C1):14–19.

Matta, V. M., R. H. Moretti, and L. M. C. Cabral. 2004. Microfiltration and reverse osmosis for clarification and concentration of acerola juice. *Journal of Food Engineering* 61(3):477–482.

Mierczynska-Vasilev, A., and P. A. Smith. 2015. Current state of knowledge and challenges in wine clarification. *Australian Journal of Grape and Wine Research* 21:615–626.

Mirsaeedghazi, H., and Z. Emam-Djomeh. 2017. Clarification of bitter orange (*Citrus aurantium*) juice using microfiltration with mixed cellulose esters membrane. *Journal of Food Processing and Preservation* 41(1).

Mirsaeedghazi, H., Z. Emam-Djomeh, S. M. Mousavi, R. Ahmadkhaniha, and A. Shafiee. 2010. Effect of membrane clarification on the physicochemical properties of pomegranate juice. *International Journal of Food Science and Technology* 45(7):1457–1463.

Mirsaeedghazi, H., S. M. Mousavi, Z. Emam-Djomeh, K. Rezaei, A. Aroujalian, and M. Navidbakhsh. 2012. Comparison between ultrafiltration and microfiltration in the clarification of pomegranate juice. *Journal of Food Process Engineering* 35(3):424–436.

APPLICATIONS OF MEMBRANE TECHNOLOGY

Mohammad, Abdul Wahab, Ching Yin Ng, Ying Pei Lim, and Gen Hong Ng. 2012. Ultrafiltration in food processing industry: Review on application, membrane fouling, and fouling control. *Food and Bioprocess Technology* 5(4):1143–1156.

Molnar, Z., S. Banvolgyi, A. Kozak, I. Kiss, E. Bekassy-Molnar, and G. Vatai. 2012. Concentration of raspberry (*Rubus idaeus* l.) juice using membrane processes. *Acta Alimentaria* 41(Supplement 1):147–159.

Mondal, M., P. P. Biswas, and S. De. 2016. Clarification and storage study of bottle gourd (*Lagenaria siceraria*) juice by hollow fiber ultrafiltration. *Food and Bioproducts Processing* 100:1–15.

Mondor, M., B. Girard, and C. Moresoli. 2000. Modeling flux behavior for membrane filtration of apple juice. *Food Research International* 33(7):539–548.

Moreno, R. M. C., R. C. de Oliveira, and S. T. D. de Barros. 2012. Comparison between microfiltration and addition of coagulating agents in the clarification of sugar cane juice. *Acta Scientiarum-Technology* 34(4):413–419.

Nandi, B. K., B. Das, and R. Uppaluri. 2012. Clarification of orange juice using ceramic membrane and evaluation of fouling mechanism. *Journal of Food Process Engineering* 35(3):403–423.

Nandi, B. K., B. Das, R. Uppaluri, and M. K. Purkait. 2009. Microfiltration of mosambi juice using low cost ceramic membrane. *Journal of Food Engineering* 95(4):597–605.

Neifar, M., R. Ellouze-Ghorbel, A. Kamoun, S. Baklouti, A. Mokni, A. Jaouani, and S. Ellouze-Chaabouni. 2011. Effective clarification of pomegranate juice using laccase treatment optimized by response surface methodology followed by ultrafiltration. *Journal of Food Process Engineering* 34(4):1199–1219.

Nourbakhsh, H., Z. Emam-Djomeh, and H. Mirsaeedghazi. 2015. Effects of operating parameters on physicochemical properties of red plum juice and permeate flux during membrane clarification. *Desalination and Water Treatment* 54(11):3094–3105.

Nourbakhsh, H., Z. Emam-Djomeh, H. Mirsaeedghazi, M. Omid, and S. Moieni. 2014. Study of different fouling mechanisms during membrane clarification of red plum juice. *International Journal of Food Science and Technology* 49(1):58–64.

Nourbakhsh, Himan, Azam Alemi, Zahra Emam-Djomeh, and Hossein Mirsaeedghazi. 2014. Effect of processing parameters on fouling resistances during microfiltration of red plum and watermelon juices: A comparative study. *Journal of Food Science and Technology* 51(1):168–172.

Onsekizoglu, Pelin, K. Savas Bahceci, and M. Jale Acar. 2010. Clarification and the concentration of apple juice using membrane processes: A comparative quality assessment. *Journal of Membrane Science* 352(1):160–165.

CLARIFICATION OF FRUIT JUICES AND WINE

Pap, N., M. Mahosenaho, E. Pongracz, H. Mikkonen, M. Jaakkola, V. Virtanen, L. Myllykoski, Z. Horvath-Hovorka, C. Hodur, G. Vatai, and R. L. Keiski. 2012. Effect of ultrafiltration on anthocyanin and flavonol content of black currant juice (*Ribes nigrum* L.). *Food and Bioprocess Technology* 5(3):921–928.

Prodanov, M., V. Vacas, A. M. Sanchez, G. L. Alonso, A. Elorza, C. Lucendo, T. Hernandez, and I. Estrella. 2013. Industrial clarification of anthocyanin-rich grape pomace extracts by crossflow membrane filtration. *International Journal of Food Science and Technology* 48(7):1426–1435.

Rai, P., and S. De. 2009. Clarification of pectin-containing juice using ultrafiltration. *Current Science* 96(10):1361–1371.

Rai, P., G. C. Majumdar, S. Das Gupta, and S. De. 2007. Effect of various pretreatment methods on permeate flux and quality during ultrafiltration of mosambi juice. *Journal of Food Engineering* 78(2):561–568.

Rai, P., G. C. Majumdar, S. Dasgupta, and S. De. 2010. Flux enhancement during ultrafiltration of depectinized mosambi (*Citrus sinensis* L Osbeck) juice. *Journal of Food Process Engineering* 33(3):554–567.

Rai, P., G. C. Majumdar, V. K. Jayanti, S. Dasgupta, and S. De. 2006. Alternative pretreatment methods to enzymatic treatment for clarification of mosambi juice using ultrafiltration. *Journal of Food Process Engineering* 29(2):202–218.

Razi, B., A. Aroujalian, A. Raisi, and M. Fathizadeh. 2011. Clarification of tomato juice by crossflow microfiltration. *International Journal of Food Science and Technology* 46(1):138–145.

Sagu, S. T., S. Karmakar, E. J. Nso, C. Kapseu, and S. De. 2014. Ultrafiltration of banana (*Musa acuminata*) juice using hollow fibers for enhanced shelf life. *Food and Bioprocess Technology* 7(9):2711–2722.

Salazar, F. N., J. P. F. de Bruijn, L. Seminario, C. Guell, and F. Lopez. 2007. Improvement of wine crossflow microfiltration by a new hybrid process. *Journal of Food Engineering* 79(4):1329–1336.

Sarkar, Biswajit, Sunando DasGupta, and Sirshendu De. 2008. Effect of electric field during gel-layer controlled ultrafiltration of synthetic and fruit juice. *Journal of Membrane Science* 307(2):268–276.

Sarkar, Biswajit, Sunando DasGupta, and Sirshendu De. 2009. Flux decline during electric field-assisted crossflow ultrafiltration of mosambi (Citrus sinensis (L.) Osbeck) juice. *Journal of Membrane Science* 331(1):75–83.

Saura, D., N. Marti, M. Valero, E. González, A. Carbonell, and J. Laencina. 2012. Separation of aromatics compounds during the clarification of lemon juice by cross-flow filtration. *Industrial Crops and Products* 36(1):543–548.

Severo, J. B., S. S. Almeida, N. Narain, R. R. Souza, J. C. C. Santana, and E. B. Tambourgi. 2007. Wine clarification from *Spondias mombin* L. pulp by hollow fiber membrane system. *Process Biochemistry* 42(11):1516–1520.

Siebert, Karl J. 1999. Effects of protein–polyphenol interactions on beverage haze, stabilization, and analysis. *Journal of Agricultural and Food Chemistry* 47(2):353–362.

APPLICATIONS OF MEMBRANE TECHNOLOGY

Siebert, Karl J., Nataliia V. Troukhanova, and Penelope Y. Lynn. 1996. Nature of polyphenol–protein interactions. *Journal of Agricultural and Food Chemistry* 44(1):80–85.

Spanos, George A., and Ronald E. Wrolstad. 1992. Phenolics of apple, pear, and white grape juices and their changes with processing and storage. A review. *Journal of Agricultural and Food Chemistry* 40(9):1478–1487.

Thakur, V., and D. K. Das Gupta. 2006. Studies on the clarification and concentration of beetroot juice. *Journal of Food Processing and Preservation* 30(2):194–207.

Toker, Ramazan, Mustafa Karhan, Nedim Tetik, Irfan Turhan, and Hatice Reyhan Oziyci. 2014. Effect of ultrafiltration and concentration processes on the physical and chemical composition of blood orange juice. *Journal of Food Processing and Preservation* 38(3):1321–1329.

Urkiaga, A., L. De las Fuentes, M. Acilu, and J. Uriarte. 2002. Membrane comparison for wine clarification by microfiltration. *Desalination* 148(1–3):115–120.

Urosevic, T., D. Povrenovic, P. Vukosavljevic, I. Urosevic, and S. Stevanovic. 2017. Recent developments in microfiltration and ultrafiltration of fruit juices. *Food and Bioproducts Processing* 106:147–161.

Ushikubo, Fernanda Yumi, Anna Paula Watanabe, and Luiz Antonio Viotto. 2006. Microfiltration of umbu (*Spondias tuberosa* Arr. Cam.) juice using polypropylene membrane. *Desalination* 200(1):549–551.

Vaillant, F., M. Cisse, M. Chaverri, A. Perez, M. Dornier, F. Viquez, and C. Dhuique-Mayer. 2005. Clarification and concentration of melon juice using membrane processes. *Innovative Food Science and Emerging Technologies* 6(2):213–220.

Vladisavljevic, G. T., P. Vukosavljevic, and M. S. Veljovic. 2013. Clarification of red raspberry juice using microfiltration with gas backwashing: A viable strategy to maximize permeate flux and minimize a loss of anthocyanins. *Food and Bioproducts Processing* 91(C4):473–480.

Wang, Fulin, and Volodymyr V. Tarabara. 2008. Pore blocking mechanisms during early stages of membrane fouling by colloids. *Journal of Colloid and Interface Science* 328(2):464–469.

Youn, Kwang-Sup, Joo-Heon Hong, Dong-Ho Bae, Seok-Joong Kim, and Soon-Dong Kim. 2004. Effective clarifying process of reconstituted apple juice using membrane filtration with filter-aid pretreatment. *Journal of Membrane Science* 228(2):179–186.

Youravong, W., Z. Y. Li, and A. Laorko. 2010. Influence of gas sparging on clarification of pineapple wine by microfiltration. *Journal of Food Engineering* 96(3):427–432.

Zarate-Rodriguez, E., E. Ortega-Rivas, and G. V. Barbosa-Canovas. 2001. Effect of membrane pore size on quality of ultrafiltered apple juice. *International Journal of Food Science and Technology* 36(6):663–667.

6

Microfiltration Techniques
Introduction, Engineering Aspects, Maintenance, and Its Application in Dairy Industries

M. Selvamuthukumaran

Contents

6.1	Introduction	156
6.2	Engineering Aspects of Microfiltration	157
6.3	Microfiltration Operational Procedure	158
6.4	Microfiltration Equipment Cleaning	158
6.5	Use of Microfiltration Techniques	159
	6.5.1 Preparation of Extended Shelf Life (ESL) Milk Products	160
	6.5.2 Cheese Processing Industries	162
	6.5.3 Preparation of Casein	162
	6.5.4 Separation of Whey Protein	162
6.6	Conclusion	163
References		163

APPLICATIONS OF MEMBRANE TECHNOLOGY

6.1 INTRODUCTION

Microfiltration is a novel membrane technique commercially applied to eliminate bacteria as well as somatic cells in skimmed milk without altering its constituents. It is a widely adopted technique in the dairy product processing industries, especially for manufacturing extended shelf life based milk products. The research studies carried out by Schmidt et al. (2012) demonstrated that microbial load could be significantly reduced from 3–6 log cfu/ml, when microfiltration techniques are used with a pore size of 1.4 µm. They also further projected that the extended shelf life milk product produced using this technique in European countries, especially Germany, Austria and Switzerland, leads to product spoilage, which can be ascribed to the survival of pathogens like *Paenibacillus* spp, *B. cereus* etc. They observed that the use of a membrane size of 1.4 µm can only eliminate vegetative cells, and does not eradicate spores of pathogenic bacteria, which can further render spoilage. If a ceramic membrane was used with a pore size of 0.8–1.4 µm, *Bacillus anthracis* was removed to the extent of a 4–5 log count (Tomasula et al., 2011).

Palmer et al. (2010) and Feng et al. (2015) reported that surface properties, i.e., surface structure, significantly helps to eradicate spore-forming bacteria. The hydrophobic and electrostatic properties may create an interaction between the membrane surface and the cells. Other important parameters, i.e., temperature, also play an important role in effectively removing the bacteria from milk using this process. The use of temperature in the range of 50–55°C can retain some pathogenic bacteria, such as the *Bacillus* and *Anoxybacillus* species (Burgess et al., 2010). Therefore, these bacteria will be recirculated within the membrane, which can cause a fouling problem for a longer time (Tang et al., 2009). Fritsch and Moraru (2008) and Walkling-Ribeiro et al. (2011) recommended the use of microfiltration at cold temperatures. The studies conducted by Griep et al. (2018) demonstrated that the spore count of *B. licheniformis* was found to be much less, i.e., 4.57, when using a 1.2-µm membrane pore size, and they attained a 2.17 log count through use of a membrane with a pore size of 1.4-µm. They further observed that the *Geobacillus* sp. was completely removed. The effect of various processing methods on the survival and stability of microorganisms and the effect of using different microfilter pore sizes on the reduction of various microorganisms are shown in Table 6.1 and 6.2. It was shown that microfiltration reduces the microbial log count to 2–3 with an enhanced stability of 1 month, and with respect to microfilter pore size the use of a membrane with a pore size of 1.4 µm significantly reduced the *Bacillus licheniformis* @ 4 logcfu/ml (Griep et al., 2018).

156

MICROFILTRATION TECHNIQUES

Table 6.1 Effect of Various Processing Methods on Survival of Microorganisms and Their Stability

Name of the Process	Microbial Log Reduction	Stability Achieved After Storing at 5–6 °C
Ultra-High-Temperature Processing	8	6 months
Pasteurization	Nil	2 weeks
Microfiltration	2–3	1 month

6.2 ENGINEERING ASPECTS OF MICROFILTRATION

Previously, microfiltration techniques were employed for substituting clarification, the centrifugal separation of whey (Merin, 1986; Merin et al., 1983; Piot et al., 1987), but the use of an existing membrane, i.e., polycarbonate and polysulfone, created issues because of its heat and chemical stability as well as its selectivity and flux. These issues were solved by using a ceramic-based membrane; this membrane is noted for its selectivity changes over time, which is found to be high, and it also causes a reduction in flux (Piot et al., 1987). The microfiltration membrane used in the dairy industry has pore sizes ranging from 20 to 0.1 µm. The membrane is composed of two sections: the macroporous support, which is the active membrane that is quoted on the surface. The macroporous material is made from either carbon or stainless steel or alumina, and its purpose is to separate the permeate substances; therefore, its pore diameter should be 10 µm. The membrane layers are made up of a mixture or combination of titanium oxide or alumina, which is further created by coating a support with a colloidal suspension of a more divided powder, which is sintered to the support through firing. The thickness of the membrane should be kept to around 3–5 µm (Grandison and Finnigan, 1996). The colloidal particle sizes can be effectively controlled to obtain a narrow pore size distribution as well as the sintering of two or more layers. The Gaussian distribution curve width restriction can significantly enhance microfiltration process selectivity. It was reported that choosing a narrow pore size distribution, i.e., 1.4 µm Sterilox, enhances the bacterial count in milk by two log cycles (Maubois, 1997). Therefore, it is advisable to use a ceramic membrane because of its mechanical resistance action, pH tolerance ranging from 1–14, easy cleaning conditions, etc. The configuration of membranes in monotubes with internal diameter ranges from 3–8 mm, with a length of around 85 cm. Many designs are available with a cylindrical channel

APPLICATIONS OF MEMBRANE TECHNOLOGY

Table 6.2 Effect of Micro Filter Pore Size on Reduction of Various Microorganisms

Name of the Microorganism Removed	Pore Size	Microbial Count Reduction Log cfu/ml
Bacillus licheniformis	1.4 µm	4
Bacillus licheniformis	1.2 µm	2.5
Geobacillus sp	1.4 µm	ND
Geobacillus sp	1.2 µm	ND

or with cross-section forms, like daisy or square or trifolium. The multichannel monoliths or monotubulars greater than 1-meter in length were assembled in bundles, which were further kept in stainless steel housings, known as MF modules. The module area may range from 0.2 to 10 m^2 or even more as per the manufacturer's design. These modules can be sterilized with steam, and they will have the retention capacity of higher heat temperatures.

6.3 MICROFILTRATION OPERATIONAL PROCEDURE

One has to take proper care while operating MF equipment so that fouling can be minimized. Fill the equipment with hot water of around 52°C and allow the air exhaust valve to open, so that the air bubbles can be easily removed. Following this, the hydraulic recirculation parameters need to be adjusted, and the physico-chemical equilibrium of milk can be re-established by slightly prewarming the milk to 50°C for a time duration of 20 minutes. The hydraulic parameters need to be readjusted during startup 10 minutes before operating in stationary conditions.

6.4 MICROFILTRATION EQUIPMENT CLEANING

Good quality water should be used for cleaning microfiltration equipment. The cleaning procedure is similar to other membrane equipment; the only variation is the type of membrane material used. To avoid bacterial contamination avoid using water containing colloidal particle suspension. Soft water should be used for cleaning the equipment, devoid of heavy mineral salts, and it should be microfiltered to satisfy this requirement.

Detergents of 1% are used for washing the equipment after hot water rinsing. Alternatively, acid cleaning can be carried out using NO3H @ 0.5% at 50°C or sanitizers like ClONa @ of 200 ppm at ambient conditions for a time duration of 15 minutes, which will give the most satisfactory results. Cleaning efficiency can be achieved by finding out the bacterial quality of last rinse water to assure the safeguard of the equipment in an efficient utilization process.

6.5 USE OF MICROFILTRATION TECHNIQUES

Microfiltration techniques are widely used in the dairy processing industries (Figure 6.1), and these are described below

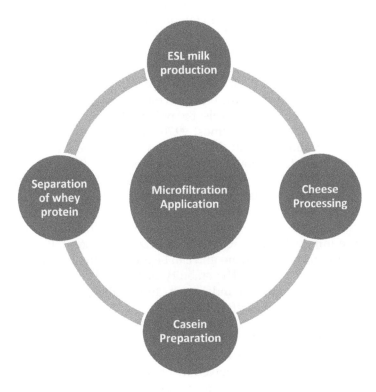

Figure 6.1 Application of microfiltration membrane techniques in dairy industries.

APPLICATIONS OF MEMBRANE TECHNOLOGY

6.5.1 Preparation of Extended Shelf Life (ESL) Milk Products

ESL milk products can be manufactured by combining a process of microfiltration cum pasteurization (Lorenzen et al., 2011) (Figure 6.2). The microbial counts are significantly reduced in ESL milk products, and they are more stable and packed under strict hygienic conditions (Rysstad and Kolstad, 2006). The microfiltration technique, a replacement for the heat preservation technique, which reduces bacteria and further enhances the safety and quality aspects of different dairy products (Pafylias et al., 1996); it also retains the taste to a greater extent. Bacterial spores can be removed from fluid milk, and spores can also be removed from whey obtained from the cheese processing industries. The orgonoleptic attributes won't be affected by this process (Meersohn, 1989). Madec et al. (1992) reported that the Listeria, salmonella and the mesophilic count could be significantly minimized with the use of a microfilter with a pore size of 1.4 μm. The microfiltration process carried out under cold atmospheric conditions reduces membrane fouling, thereby further spore germination can be prevented (Fristch and Moraru, 2008). Bacteria-free milk can be obtained by altering the design, structure and composition of the membranes (Malmberg and Holms, 1988; Olesen and Jensen, 1989). Hoffmann et al. (2006) reported that pathogenic bacteria could be successfully eliminated without altering the protein levels; the only protein reduction was noted to be around 0.02 to 0.03%. Almost 99.70% of skimmed milk's bacterial load can be minimized with the provision of a filtrate circulatory system in a concentrate circulatory system of membrane filters (Olesen and Jensen, 1989). The Alfa laval Tetra Alcross Bactocath eliminates bacterial spores from milk by fractionating the fluid milk into skimmed milk and cream; the microfiltration process further separates the bacterial spores from skimmed milk, retaining bacterial spores, which accounts for 0.5% of the true milk volume. After the pasteurization process, the retentate can be incorporated with cream and can be mixed with filtered milk after the pasteurization process. The stability of ESL milk can be enhanced to 45 days at 3–4°C (Olesen and Jensen, 1989; Puhan, 1992; Saboya and Maubois, 2000; Goff and Griffiths, 2006).

The integrated approach of microfiltration cum heat treatment reduces the somatic cells up to 100% (Pedersen, 1992). The stability of refrigerated milk was extended to 18 days at 7°C with this combined effort (Damerow, 1989). Stack and Sillen (1998) showed that microorganisms, especially bacteria and spores, could be effectively removed by using the microfiltration process, when compared to bactofugation (Stack and Sillen, 1998).

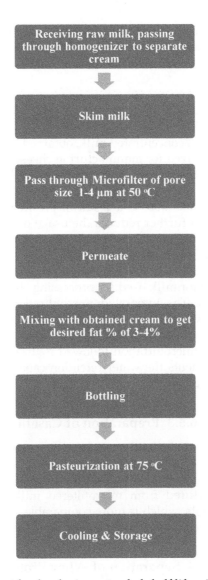

Figure 6.2 Flowsheet for developing extended shelf life milk (ESL).

APPLICATIONS OF MEMBRANE TECHNOLOGY

Brans et al. (2004) reported that liquids could be easily passed through a membrane made up of silicon nitride with narrow pore sizes. It also further minimizes microorganisms and fouling in milk (Van Rijn and Kromkamp, 2001).

6.5.2 Cheese Processing Industries

The protein, i.e., casein concentrated milk, obtained by microfiltration can be very efficiently utilized for manufacturing cheese, and this is ascribed to the removal of pathogenic microorganisms as well as major milk component optimization. The curd firmness is improved, and ripening processes are highly accelerated as a result of pre-treatment of milk with microfiltration, which further reduces the usage of additives, and facilitates heat transfer at higher temperatures (Pierre et al., 1992; Caron et al., 1997; Maubois, 2002; Schafroth et al., 2005).

Mistry and Maubois (1993) reported that protein standardization seems to be possible for milk used for processing cheese by using microfiltration techniques. The integrated microfiltration cum high-temperature processing is very effective for killing microbes compared to the bactofugation process alone. Coagulation of rennet is modified when milk is subjected to microfiltration followed by heat treatment and also simultaneously augments the water retention capacity of the coagulum (Pedersen and Ottosen, 1992).

6.5.3 Preparation of Casein

The β, α and к-casein can be fractionated and concentrated at specific ionic strength under 4.2–4.6 pH using microfiltration techniques (Famelart et al., 1989; Murphy and Fox, 1990). Papadatos et al. (2003) observed that the cheese manufactured from microfiltered milk in North America yielded higher outputs, which is more comparable, i.e., 2–3 times higher, than traditional processing without using a membrane.

6.5.4 Separation of Whey Protein

Microfiltration techniques were used for purifying and concentrating proteins from skimmed milk whey (Govindasamy-Lucey et al., 2007; Lawrence et al., 2008). The immunoglobulins were separated from cholesterol using this technique (Regester et al., 2003; Piot et al., 2004).

162

6.6 CONCLUSION

A ceramic membrane with a pore size of 1.2–1.4 μm can be used for producing bacterial free milk or milk with a reduced log count of microorganisms. The β, α and κ-casein can be fractionated and concentrated at specific ionic strengths by using microfiltration techniques. The proteins can be purified and concentrated from skimmed milk whey; thus, the extended shelf life product can be prepared successfully using this microfiltration technique.

REFERENCES

Brans, G., C. G. P. H. Schroen, R. G. M. Van der Sman, and R. M. Boom. 2004. Membrane fractionation of milk: State of the art and challenges. *J. Memb. Sci.* 243(1–2):263–272.

Burgess, S. A., D. Lindsay, and S. H. Flint. 2010. Thermophilic bacilli and their importance in dairy processing. *Int. J. Food Microbiol.* 144(2):215–225.

Caron, A., D. Saint Gelais, and Y. Pouliot. 1997. Coagulation of milk enriched with ultrafiltered or diafiltered microfiltered milk retentate powders. *Int. Dairy J.* 7(6–7):445–451.

Damerow, G. 1989. Die Anwendung der mikrofiltration fur die konsummilch, Kessel-Milch, Molke. *Dtsch. Molkerei Zeitung* 110:1602–1608.

Famelart, M. H., C. Hardy, and G. Brule. 1989. Optimisation of the preparation of p-casein-enriched solution. *Lait* 69(1):47–57.

Feng, G., Y. Cheng, S. Y. Wang, D. A. Borca-Tasciuc, R. W. Worobo, and C. I. Moraru. 2015. Bacterial attachment and biofilm formation on surfaces are reduced by small-diameter nanoscale pores: How small is small enough? *NPJ Biofilms Microbiomes* 1:15022.

Fritsch, J. and C. I. Moraru. 2008. Development and optimization of a carbon dioxide-aided cold microfiltration process for the physical removal of microorganisms and somatic cells from skim milk. *J. Dairy Sci.* 91(10):3744–3760.

Goff, H. D. and M. W. Griffiths. 2006. Major advances in fresh milk and milk products: Fluid milk products and dairy desserts. *J. Dairy Sci.* 89(4):1163–1173.

Govindasamy-Lucey, S., J. J. Jaeggi, M. E. Johnson, T. Wang, and J. A. Lucey. 2007. Use of cold microfiltration retentates produced with polymeric membranes for standardization of milks for manufacture of pizza cheese. *J. Dairy Sci.* 90(10):4552–4568.

Grandison, A. S. and T. J. A. Finnigan. 1996. Microfiltration. In: A. S. Grandison, Lexis, M. J. (Eds.), *Separation Process in the Food and Biotechnology Industries*, Woodhead Publ. Ltd.: Cambridge, England, pp. 141–153.

Griep, E. R., Y. Cheng, and C. I. Moraru. 2018. Efficient removal of spores from skim milk using cold microfiltration: Spore size and surface property considerations. *J. Dairy Sci.* 101(11):9703–9713.

APPLICATIONS OF MEMBRANE TECHNOLOGY

Lawrence, N. D., S. E. Kentish, A. J. OConnor, A. R. Barber, and G. W. Stevens. 2008. Microfiltration of skim milk using polymeric membranes for casein concentrate manufacture. *Sep. Purif. Technol.* 60(3):237–244.

Lorenzen, P. C., I. C. Decker, K. Einhoff, P. Hammer, R. Hartmann, W. Hoffmann, D. Martin, J. Molkentin, H. G. Walte, and M. Devrese. 2011. A survey of the quality of extended shelf life (ESL) milk in relation to HTST and UHT milk. *Int. J. Dairy Technol.* 64(2):166–178.

Madec, M. N., S. Mejean, and J. L. Maubois. 1992. Retention of Listeria and Salmonella cells contaminating skim milk by tangential membrane microfiltration (Bactocatch process). *Lait* 72(3):327–332.

Malmberg, R. and S. Holms. 1988. Producing low-bacteria milk by microfiltration. *N. Eur. Food Dairy J.* 54:30–32.

Maubois, J.-L. 1997. Current uses and future perspectives of MF technology in the dairy industry. *Bull. Int. Dairy Fed.* 320:37–40.

Maubois, J. L. 2002. Membrane microfiltration: A tool for a new approach in dairy technology. *Aust. J. Dairy Technol.* 57:92–96.

Meersohn, M. 1989. Nitrate-free cheese making with the Bactocatch. North Eur. *Food Dairy J.* 55:108–113.

Merin, U. 1986. Bacteriological aspects of microfiltration of cheese whey. *J. Dairy Sci.* 69(2):326–328.

Merin, U., S. Gordin, and G. B. Tanny. 1983. Microfiltration of sweet cheese whey. *J. Dairy Res.* 18:153–160.

Mistry, V. V. and J. L. Maubois. 1993. Application of membrane separation technology to cheese production. In: P. F. Fox (Ed.), *Cheese: Chemistry, Physics and Microbiology*, Chapman & Hall: London, pp. 493–522.

Murphy, J. M. and P. F. Fox. 1990. Fractionation of sodium caseinate by ultrafiltration. *Food Chem.* 39(1):27–38.

Olesen, N. and F. Jensen. 1989. Microfiltration. The influence of operating parameters on the process. *Milchwissenschaft* 44:476–479.

Pafylias, I., M. Cheryan, M. A. Mehaiab, and N. Saglam. 1996. Microfiltration of milk with ceramic membranes. *Food Res. Intl.* 29(2):141–146.

Palmer, J. S., S. H. Flint, J. Schmid, and J. D. Brooks. 2010. The role of surface charge and hydrophobicity in the attachment of *Anoxybacillus flavithermus* isolated from milk powder. *J. Ind. Microbiol.Biotechnol.* 37(11):1111–1119.

Pedersen, P. J. and N. Ottosen. 1992. New applications of membrane processes. IDF Spl Iss 9201. Brussels, Belgium, pp. 67–76.

Papadatos, A., M. Neocleous, A. M. Berger, and D. M. Barbano. 2003. Economic feasibility evaluation of microfiltration of milk prior to cheesemaking. *J. Dairy Sci.* 86(5):1564–1577.

Pierre, A., J. Fauquant, Y. Le Graet, M. Piot, and J. L. Maubois. 1992. Préparation de phosphocaseinate natif par microfiltration sur membrane. *Lait* 72(5):461–474.

Piot, M., J. Fauquant, M. N. Madec, and J. L. Maubois. 2004. Preparation of serocolostrum by membrane microfiltration. *Lait* 84(4):333–341.

Piot, M., J.-L. Maubois, P. Schaegis, and R. Veyre. 1984. Microfiltration en flux tangentiel des lactosérums de fromagerie. *Lait* 64:102–120.

164

Piot, M., J. C. Vachot, M. Veaux, J.-L. Maubois, and G. E. Brinkman. 1987 Écrémage et épuration bactérienne du lait entier cru par microfiltration sur membrane en flux tangentiel. *Tech. Lait Market.* 1016:42–46.

Puhan, Z. 1992. *New Applications of Membrane Processes.* IDF Spl Iss 9201. Brussels, Belgium, pp. 23–32.

Regester, G. O., D. A. Belford, R. J. West, and C. Goddard. 2003. Development of minor dairy components as therapeutic agents-Whey growth factor extract, a case study. *Aust. J. Dairy Technol.* 58:104–106.

Rysstad, R. and J. Kolstad. 2006. Extended shelf life milk-advances in technology. *Int. J. Dairy Technol.* 59(2):85–96.

Saboya, L. V. and J. L. Maubois. 2000. Current developments of microfiltration technology in the dairy industry. *Lait* 80(6):541–553.

Schafroth, K., C. Fragnière, and H. P. Bachmann. 2005. Herstellung von Käse aus microfiltrierter. konzentrierter Milch. *Deutsche Milchwirtschaft* 56:861–863.

Schmidt, V. S. J., V. Kaufmann, U. Kulozik, S. Scherer, and M. Wenning. 2012. Microbial biodiversity, quality and shelf life of microfiltered and pasteurized extended shelf life (ESL) milk from Germany,Austria and Switzerland. *Int. J. Food Microbiol.* 154(1–2):1–9.

Stack, A. and G. Sillen. 1998. Bactofugation of liquid milks. *Nutr. Food Sci.* 98(5):280–282.

Tang, X., S. H. Flint, J. D. Brooks, and R. J. Bennett. 2009. Factors affecting the attachment of microorganisms isolated from ultrafiltration and reverse osmosis membranes in dairy processing plants. *J. Appl. Microbiol.* 107(2):443–451.

Tomasula, P. M., S. Mukhopadhyay, N. Datta, A. Porto-Fett, J. E.Call, J. B. Luchansky, J. Renye, and M. Tunick. 2011. Pilot-scale crossflow-microfiltration and pasteurization to remove spores of *Bacillus anthracis* (Sterne) from milk. *J. Dairy Sci.* 94(9):4277–4291.

Van Rijn, C. J. and J. Kromkamp. 2001. Method for filtering milk. WO Patent 0209527.

Walkling-Ribeiro, M., O. Rodriguez-Gonzalez, S. Jayaram, and M. W.Griffiths. 2011. Microbial inactivation and shelf life comparison of 'cold' hurdle processing with pulsed electric fields and microfiltration,and conventional thermal pasteurisation in skim milk. *Int. J. Food Microbiol.* 144(3):379–386.

7

Applications of Membrane Technology in Whey Processing

Kirty Pant, Mamta Thakur and Vikas Nanda

Contents

7.1	Introduction to Membrane Technology	168
7.2	Membrane Technology in Dairy Industry: An Overview	169
	7.2.1 Concentration and Component Separation	169
	7.2.2 Eradication of Bacteria to Extend Shelf Life	172
	7.2.3 Wastewater Treatment	173
	7.2.4 Utilization of Electrodialysis and Pervaporation in Dairy Industry	174
7.3	Whey: Chemical Composition and Properties	175
7.4	Whey Protein Recovery and Concentration	177
	7.4.1 Development of Whey Protein Concentration and Isolates	180
	7.4.2 Recovery of Whey Protein Fractionations	181
7.5	Concentration and Purification of Lactose	184
7.6	Whey Demineralization (Desalination)	185
7.7	Challenges Related to Membrane Processing	188
	7.7.1 Concentration Polarization	189
	7.7.2 Membrane Fouling	189
7.8	Future Trends	191
7.9	Conclusion	192
References		192

APPLICATIONS OF MEMBRANE TECHNOLOGY

7.1 INTRODUCTION TO MEMBRANE TECHNOLOGY

"Membrane technology" is the comprehensive terminology which represents processes of separation and filtration using specific screens with semi-permeable characteristics to obtain two liquids possessing distinct characteristics from a common feed stream by selectively allowing some components to move while restricting others (Nissar et al., 2018; Chen et al., 2018). The application of membrane technology as a processing and separation technique is gaining popularity commercially, especially in the food industry. Membrane separation is a novel technology used for processing new compounds and foods, and also used as a substitute method to conventional techniques (Le et al., 2014). The liquid which is being restricted by the membrane filters is termed the "concentrate" or "retentate," while the liquid which is able to move through the membrane filter is termed the "permeate." The "transmembrane pressure" (also known as "hydrostatic pressure gradients") throughout the membrane and the concentration gradients are the principle factors, governing the efficiency of the membranes; the electric potential also plays a significant role in a few cases (Winston and Sirkar, 1992). In various processing industries, membrane separation techniques are broadly used to separate and to purify specific components from the rest of the fluid mixture. In this technique the target constituent might be the desired or undesired element, segregated in order to increase the purity of the native mixture (Mohsenin, 1980; Lewis, 1990; Mulder, 1991) Membrane separation techniques are cost-effective and environmental friendly; therefore, they are also considered as green technology (Chen et al., 2018). In several cases, the application of membrane technology is more beneficial than traditional methods. For example, using a suitable membrane is highly economical instead of applying high-temperature treatments for the elimination of micro-organisms during sterilization and cold pasteurization processes for reducing energy consumption. Application of membrane filtration to eradicate micro-organisms for the extension of the shelf life of food products, rather than using preservatives and additives, builds an environmentally friendly image for processed foods and food processing practices. Concentration by membrane filtration has an advantage over thermal evaporation because it preserves the natural taste of the food material and the nutritive value of heat-sensitive components. Presently the most active and advantageous features of membrane technology are wastewater treatment and the recovery of valuable constituents from diluted effluents (Le et al., 2014).

Numerous separation technologies exist but membrane separation technology has caused a significant transformation in dairy food processing by contributing many benefits. These include component separation at lower temperatures, separating a component in its native form, the removal of bacteria, filtration, wastewater treatment and being energy-efficient, etc. (Marella et al., 2013; Chen et al., 2018).

7.2 MEMBRANE TECHNOLOGY IN DAIRY INDUSTRY: AN OVERVIEW

Membrane utilization in the dairy industry is appreciably increased with the application of novel based material such as polyamides, cellulose acetate and polysulphones with the combination of advanced technological processes like nanofiltration (NF), diafiltration (DF) and reverse osmosis (RO), as given in Table 7.1. The current scenario in the dairy sector is that the kinds of membrane technologies being used are nano-filtration, micro-filtration, ultra-filtration and reverse osmosis (Vourch et al., 2005; Balannec et al., 2005).

7.2.1 Concentration and Component Separation

Milk concentration is the process by which water is removed from milk to decrease the cost of the packaging material, storage and transportation of milk and milk products. Traditionally heat exchange principle-based methods, such as flash (Kiesner et al., 1994) and falling film evaporation (Liu, 2010), have been used to concentrate the milk. However, these techniques may influence the configuration, rheological properties and heat stability of the product, and they are also highly energy-consuming, and, consequently, the characteristics of the end products are influenced (Liu et al., 2017). Membrane separations and phase separation technologies sound similar, but they are different from each other, as shown in Figure 7.1. These technologies are possessed at a lower cost operation, they are environment friendly, and they are comprised of simple process structures (Kolfschoten et al., 2011). Many researchers have reported that the application of ultrafiltration and reverse osmosis in the pre-concentration of soft cheese, quarg and yogurt, respectively, significantly improved the quality of the final cheese (Jevons, 1997). Apart from milk concentration, membrane separation processes are also being applied to isolate various milk constituents, such as lactose, minerals, whey proteins and

APPLICATIONS OF MEMBRANE TECHNOLOGY

Table 7.1 Characteristics of Commonly Used Membrane Techniques in Dairy Sector (Winston and Sirkar, 1992; Chen et al., 2018)

Parameter	Microfiltration	Ultrafiltration	Nanofiltration	Reverse Osmosis
Pore size (µm)	0.1–1	0.01–0.1	0.001–0.01	0.0001–0.001
Working Principle	Sieving	Sieving	Donnan effect, Dissolving diffusion	Dissolving diffusion
Pressure (Bar)	<1 (Low pressure)	1–10 (Medium Pressure)	10–40 (Medium to High Pressure)	30–100 (High Pressure)
MWCO (KDa)	>200	1–200	0.3–1	0.1
Filtration Module	Tubular, ceramic-based	Tubular, spiral-wound, plate and frame, hollow-fiber, ceramic-based, polymer-based	Spiral- wound	Spiral-wound, plate and frame, polymer-based
Size of intercept components	0.1–20 µm	5–100 nm	>1 nm	0.1–1 nm
Retentate compounds	Comparatively low retentate-fat globules, protein separation, bacterial cells and other particulates	Large retentate- casein micelles, fat globules, mineral colloids, somatic cells and microbial cells	Separate lactose, monovalent salts and water, comparatively low productivity	Lactose and divalent salts, comparatively low productivity
Applications	Filtration, separation, clarification, elimination of bacterial cells	Grading, concentration, refining of macromolecules	Isolation, purification, and Food process enrichment, Pharmaceutical and biochemical industries	Removal of salt from the aqueous medium, concentration of lower molecular weight compounds

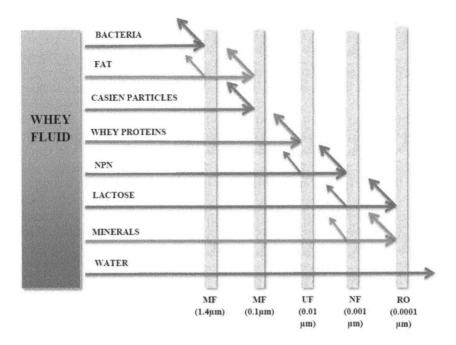

Figure 7.1 A systematic diagram of whey processing using membrane technology.

saccharides. Milk proteins consist of casein micelles (diameter < 200nm) and whey proteins (diameter < 20nm) that can be separated via membrane filters with pore size ranges from 0.05 to 0.2 nm diameter (Heino et al., 2007).

Whey protein is comprised of β-lactoglobulin, α-lactalbumin, lactoferrin and immunoglobulins that exhibit immune-enhancing properties (Marshall, 2004). Samtlebe et al. (2015) reported nearly 40% β-lactoglobulin and 60% α-lactalbumin can be separated from the whey using permeation and phage retention techniques combined with a simultaneous filtration process. However, Al-akoum et al. (2002) reported that the application of vibratory shear improved processing to yield approximately 65% α-lactalbumin and 25–30% β-lactoglobulin with a bit of higher transmission rates. Rotating ceramic membrane techniques are more appropriate since they provide a better balance in the flux and whey protein transmission (Espina et al., 2009).

APPLICATIONS OF MEMBRANE TECHNOLOGY

In comparison to traditional techniques such as rennet coagulation and acidification, the micellar structure of casein is not affected by membrane separation processes (Lawrence et al., 2008). The temperature during membrane filtration processes range from 45 to 50°C, which is favorable for regulating high flux and also to resist the growth of mesophilic bacteria (Hurt et al., 2010). Low temperature is required to isolate β-casein as it exists in the serum phase; thus, new separation techniques have been established, such as polyvinylidene fluoride (PVDF) and polyethersulfone (PES) membrane, that are used in the enrichment of β-casein at low temperatures (< 20°C) so that the physicochemical properties, composition and protein profile remain same in the final casein. Moreover, the utilization of a PSE membrane increases the flux rate and suppresses the fouling issue (Crowley et al., 2015). Chai et al. (2017) observed that membrane filtration with a transverse vibrating system of PVDF membranes with a pore size of 0.04 μm operating at 10°C does well in the concentration and separation of milk protein, and the native structure of the obtained protein is preserved with a better shelf life.

Protein-enriched yogurt is produced from fractionate skimmed milk by applying the combined effect of microfiltration and ultrafiltration techniques, which separates the lactose content of end product, reducing it up to 50% (Rinaldoni et al., 2009). Morr and Brandon (2008) assessed the combined effect of microfiltration and ultrafiltration over fractionate lactose and sodium content from skimmed milk, and results showed that the 90–95% of lactose and sodium bifurcated without influencing consumer acceptance regarding product appearance and flavor. When an ultrafiltration membrane is employed to separate lactose from goat's milk, some of the specific components such as casein, serum proteins and fat globules are kept, though this issue can be rectified by optimizing parameters like cross-flow velocity and transmembrane pressure (Abidin et al., 2014).

7.2.2 Eradication of Bacteria to Extend Shelf Life

Milk is comprised of an abundance of particles with diverse size ranges and characteristics, such as fat globules (0.2–15 um), casein micelles (0.03–0.3 um), somatic cells (6–15 um) and bacteria (0.2–6 um) (Saboya et al., 2000). Somatic cells and microbial load in milk influences the quality and flavor, and also decreases the shelf life of end products. Traditional methods are based on heat treatment given to milk in order to inactivate the microbial cells to promote the shelf life; however, thermal processing also affects the nutritional value as well as the flavor profile of the final product (Hinrichs et al., 2011). Membrane separation systems function at low temperatures, which eliminate bacteria efficiently without influencing the nutritional value and flavor profile of the end product. It also decreases the cost of processing and transportation; thus, microfiltration is broadly used for the elimination of bacteria (Solanki et al., 2001; Fristch and Moraru, 2008; Madaeni et al., 2011). Treatment given by cold MF could also diminish the possibilities of membrane fouling caused by micro-organisms, and it also avoids germination of thermophilic spores (Fristch and Moraru, 2008). Chemical and microbiological studies have been conducted to evaluate the MF effect on skimmed milk composition, and a drastic improvement was also observed in the permeate flux by increasing the velocity from 5 to 7 m/s. Recently the cross-flow microfiltration (CFMF) technique has emerged as one of the new advanced industrial separation methods in the dairy industry (Fristch and Moraru, 2008; Tomasula et al., 2011; Fauquan et al., 2014; Schmitd et al., 2012). Sterilization performed with inorganic ceramic membranes are used to retain the flavor of milk; it also extends the shelf life of the final product. This processing technology is integrated with a slight thermal treatment with cold pasteurization (Lorenzen et al., 2011), and the end product is known as extended shelf life (ESL) milk (Hoffmann et al., 2006).

The MF technique is used to develop ESL milk in which bacterial cells are eliminated from the final product by avoiding compositional alterations and insignificant reductions (approximately 0.02 to 0.03%) in the total protein content (Hoffmann et al., 2006). ESL milk has a prolonged shelf life of approximately 3 weeks, more than HTST-pasteurized milk (typically 10 days), and sensory profile studies have reported that there is no significant variation between ESL milk and pasteurized milk throughout storage.

7.2.3 Wastewater Treatment

The dairy industry is one of the food industries known for its huge water consumption, which leads to pollution because of the dispersion of large amounts of wastewater (Le et al., 2014; Rahimi et al., 2016). The application of membrane separation technology is one of the best and active approaches to treat wastewater (Le et al., 2014). Wastewater produced from the dairy industry carries huge amounts of organic matter and nutrients, especially lactose which is capable of extents the chemical and biochemical oxygen demands and causes pollution. To prevent this, it is essential to implement membrane separation techniques, which play

APPLICATIONS OF MEMBRANE TECHNOLOGY

a significant part in secondary biological treatment as this technique is energy-saving and can contribute to wastewater zero-emission (Cuartas-Uribe et al., 2009; Luo et al., 2010; Andrade et al., 2014). Numerous studies have been reported concerning lactose recovery via ultrafiltration, nano-filtration and reverse osmosis techniques (Pouliot et al., 1999; Barba et al., 2000; Barba et al., 2001). When compared to ultrafiltration, nanofiltration and reverse osmosis methods possess high efficiency for lactose recovery but operate at higher pressures (Cheang et al., 2004; Atra et al., 2005; Barba et al., 2002). Researchers found that the lactose recovery rate in permeate was improved by applying membranes with molar weights of 3, 5 and 10 kDa, which produced 70–80%, 90–95% and 100% lactose, respectively (Chollangi and Hossain, 2007). In addition, lactose treated with β-galactosidase enzymes and put through a hydrolysis process produces glucose and galactose in continuous stirred tank-ultrafiltration (CSTR-UF) (Namvar-Mahboub et al., 2012). Human milk oligosaccharides (HMOs) show a significant effect on the growth and overall development of infants and act as a functional food ingredient (Hickey et al., 2012). Whereas, animal milk also comprises oligosaccharides, which possess similar characteristics as human milk oligosaccharides (Oliveira et al., 2015). Luo et al. (2014) described that N-acetylneuraminic acid (sialic acid) is sialyllactose bound to β-lactose and showed a high level of permeation of 3'-sialyl-lactose, which is achieved by utilizing a combined UF/NF membrane structure for the valorization and value addition of dairy by-products by incorporating engineered sialidase.

7.2.4 Utilization of Electrodialysis and Pervaporation in Dairy Industry

Besides the above-discussed membrane separation technologies, electro-dialysis and pervaporation techniques are also applied in dairy industry processing. Electrodialysis is a process that is applied for the concentration or separation of ions in a solution, and principally depends on selective electromigration through semi-permeable membranes under the effect of the potential gradient (Chen et al., 2018). Nowadays, this process is extensively used in the dairy industry and has been effectively applied for the demineralization of skimmed milk and whey (Andrés et al., 1995). Researchers found that the application of electrodialysis for the demineralization of skimmed milk (rate up to 75%) was much better than the previous methods with demineralization rate ranges from 30–40% (Bazinet et al., 2001). Chen et al. (2016) observed that electrodialysis efficiently

APPLICATIONS OF MEMBRANE TECHNOLOGY IN WHEY PROCESSING

eliminated the lactate ions from acid whey in order to resolve unit operational difficulties in downstream spray-drying operations. Whereas, electrodialysis has been efficaciously proven for the recovery of lactic acid from fermentation broths (Wee et al., 2015), and this procedure is also used in the demineralization of sweet whey earlier in the whey powder production process (Greiter et al., 2002). However, applications of electrodialysis are still in their beginning stages, and its prospective capabilities have not yet been entirely exploited (Fidaleo et al., 2006). The prevaporation process is a selective membrane separation technique, where the liquid feed flows on one side of the membrane, and the concentrated feed components are obtained in vapor form from another side of the membrane (Marella et al., 2013). It is the only membrane separation technique where the phase transition takes place with the feed in a liquid state and the permeate in a vapor state, which is feasible by regulating the partial vacuum on the permeate side of the membrane (Marella et al., 2013). It can also be applied to concentrate particular compounds in a liquid mixture. In hydrophobic pervaporation, the volatile hydrophobic compounds such as flavor pass from the polymeric membrane more efficiently than when compared to water, and thus it is concentrated in the permeate (Lipnizki et al., 1999). In earlier reports, it was applied to concentrate acids, esters and ketones in model flavor mixtures, and the features of the feed mixture (dairy ingredients and pH) were observed to change the pervaporation function of the flavor compounds (Overington et al., 2011).

7.3 WHEY: CHEMICAL COMPOSITION AND PROPERTIES

Whey makes up 80–90% of the volume of milk introduced in the process, and it comprises approximately 50% of the total nutrients that exist in the original milk, such as vitamin, mineral, lactose and soluble protein, as shown in Table 7.2 (Bylund, 2003). Whey is known as a by-product of the cheese industry, obtained as a yellowish-green liquid fraction after collecting curd precipitates during the production of cheese, and its composition is highly influenced by the property of milk as well as the technology applied (Hinkova et al., 2012). Milk treated with rennet (containing protease chymosin) produces sweet whey with a pH range from 5.9–6.6, whereas casein coagulates obtained by the addition of mineral acid (hydrochloric or sulphuric acid), organic acid (lactic acid) or by treating with *lactobacilli*, leads to the production of acid whey, with a pH range of 4.3–4.6 (Bylund, 2003). Lactose, present in high quantities in whey, is comprised of almost

APPLICATIONS OF MEMBRANE TECHNOLOGY

Table 7.2 Approximate Composition of Sweet and Acid Whey in Percentage

Component	Sweet whey (%)	Acid whey (%)
Moisture content	93.6	93.5
Total solids	6.4	6.5
Fat	0.05	0.04
Lactose	4.8	4.9
NPN(non- protein nitrogen)	0.18	0.18
True protein content	0.58	0.58
β-Lactoglobulin (β-LG)	0.32	0.32
α-Lactalbumin (α-LA)	0.12	0.12
Bovine serum albumin (BSA)	0.04	0.04
Immunoglobulins(Igs)	0.08	0.08
Lactoferrin(LF)	0.02	0.02
Lactoperoxidase(LP)	0.03	0.003
Enzymes	0.003	0.003
Ash content	0.5	0.8
Ca	0.043	0.12
K	0.16	0.16
Na	0.050	0.050
P	0.040	0.065
Cl	0.11	0.11
Lactic acid	0.05	0.4

(Bylund, 2003; Tunick, 2019)

80% of dry matter and approximately 10% proteins and other constituents such as non-protein nitrogenous compounds, fats, acids, minerals and vitamins (water soluble) (Hinkova et al., 2012). α-lactoalbumin and β-lactoglobulin are the major protein fractions present in whey which contributes 70% of the total whey proteins, and they are mainly responsible for the various functional properties driven by whey such as the gelation, hydration, foaming and emulsifying characteristics of the whey solution, which determines that whey is a good source of raw material as a sugar substitute in the manufacturing of ice cream or yogurt. However, whey contains large amounts of mineral salt, especially sodium chloride (NaCl), which leads to poor quality whey (Roman et al., 2009).

Liquid whey is considered as one of the largest reservoirs for producing quality protein food; it's still out of the reach of human consumption networks (Richards et al., 2002; Yorgun et al., 2008). The worldwide production of whey is estimated to be approximately 190 million tons per year; however, 50% of this volume is processed, whereas the rest goes into sewage and other natural water bodies (Richards et al., 2002; Yorgun et al., 2008). Whey as a waste is considered as a serious pollutant because it causes severe environmental damage by imposing a very high chemical oxygen demand (COD) of 60,000–80,000 mg/L and biological oxygen demand (BOD) of 30,000–50,000 mg/L (Macwan et al., 2016). The disposal of whey also accounts for a significant loss of energy and essential nutrients and leads to a waste load equivalent to approximately 100–175 times its domestic water waste (Smithers, 2008).

Various technologies introduced in whey processing, especially membrane technology, in order to create numerous valuable by-products from whey, with various applications in pharmaceutical and food industries for promoting health and combat protein shortage (Figure 7.2).

7.4 WHEY PROTEIN RECOVERY AND CONCENTRATION

Traditionally isolation of whey proteins was carried out by utilizing different precipitation techniques, whereas the heat denaturation process was extensively applied to concentrate and to separate whey proteins from whey serum. But the problem associated with the heat-precipitated whey protein (HPWP) was that the protein coagulates shows either sparingly soluble or insoluble behavior based on the conditions dominant during denaturation (Bylund, 2003). Since the late 20th century, the extensive acceptance of pressure-driven membrane techniques such as ultrafiltration (UF) has signified a lead toward a major technological development in the dairy industry for facilitating whey protein recovery in its native state from whey serum (Price, 2019). The ultrafiltration technique is one of the common processes used in dairy sectors in order to concentrate dairy protein for development of an end product containing a high amount of dairy proteins, such as whey protein concentrates (WPC), whey protein isolates (WPI) and milk protein concentrates (MPC). Approximately 350,000 m^2 of the UF membrane has been installed in the dairy sector since

APPLICATIONS OF MEMBRANE TECHNOLOGY

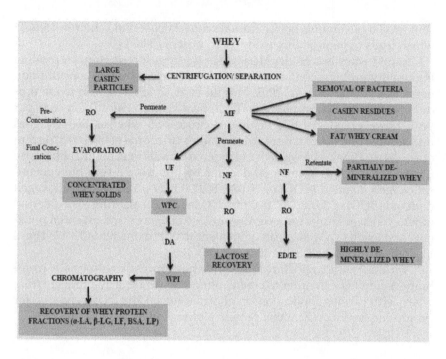

Figure 7.2 A systematic flow chart showing the recovery of various whey products via membrane processing.(MF- microfiltration; UF- ultrafiltration; NF- nanofiltration; RO- reverse osmosis; DA; diafiltration; WPC- whey protein concentrates; WPI- whey protein isolate; ED- electrodialysis; IE- ion-exchange; α-LA- α-Lactalbumin; β-LG- β-Lactoglobulin; BSA- Bovine serum albumin; LF- Lactoferrin; LP- Lactoperoxidase).

2007 (Gésan-Guiziou et al., 2007). Whey protein concentration through the ultrafiltration membrane technique is based on the principle of selective permeation phenomena, i.e., the permeation of tiny molecules such as water, non-protein nitrogenous molecules, lactose, and mineral, while efficiently retaining the whey protein molecules (Etzel and Arunkumar, 2015). UF membranes comprise a broad range of molecular weight cutoffs (MWCOs) spread confined in the membrane separation spectrum. Despite this, in the dairy industry the isolation of whey protein is accomplished through the utilization of a 10,000 Dalton (10 kDa) MWCO UF membrane in combination with other processors, though nowadays a tighter 5 kDa UF membrane has been introduced in multistage dairy plants for advanced concentration (Kelly, 2019). Numerous researchers have described various

APPLICATIONS OF MEMBRANE TECHNOLOGY IN WHEY PROCESSING

ways to recover and concentrate whey proteins. Rinn et al. (1990) defined a method to produce WPC through applying a 10 kDa ultrafiltration membrane, which implicates a 24-fold volume deduction combined with three diafiltration volumes (pp of initial concentrate of deionized water). Barba et al. (2001) produced WPC with a two-stage diafiltration process; however, Rektor and Vatai (2004) reported the application of UF membranes with large pore sizes of 100 kDa for the filtration of mozzarella whey, and the yield obtained was only 75%. The implementation of the diafiltration technique always helps to promote membrane flux and enhance protein concentration in the dried matter. The significance of diafiltration is to maximize the protein content on a dry basis, which is generally over 80%; whereas, without application of diafiltration, the protein concentration only reaches about 60–70%. Hence, protein concentration ranges from 20 to 80%, which can be efficiently acquired by a combination of UF and diafiltration based on the specific requirement (Wang and Guo, 2019). The degree of protein concentration from WPC35 (35%) to around WPC80 (~80%), which is directly influenced by the volume concentration factor (VCF), for example, WPC35, is obtained when VCF reaches 4.5–7.0, whereas for WPC60, a VCF of 13–20 is required. The combination of an UF membrane with a diafiltration technique helps to facilitate an additional permeation of non-protein components from whey; hence, it promotes enrichment of WPC by ~85% through developing a range of VCF of 30–35 (Kelly, 2019). After the whey is subjected to a centrifugal cream separator machine, it still contains leftover fat residues, which usually creates issues in confining the higher protein concentration of WPC up to 80% or more, as the fat residues are also retained by the UF membrane with the protein molecules. During the production of WPC80%, an immense amount of protein concentration takes place, but approximately 7–8% of the fat is also retained in the end product (powder). Moreover, the retained fat residue is susceptible to oxidation and adversely affects the sensory attributes of protein ingredients in the final product (Kelly, 2019). The development of higher protein concentrated products has only been attained using the application of an ion-exchange resin system, wherein it has played a significant role for achieving protein concentration beyond 90%; hence, whey protein isolates (powder) are produced with the Vistec process, whereas later on it's known as bio-isolates. Afterward, the microfiltration membrane system became established in the dairy sector and has now become mainstream; it is recognized as giving a better flux rate in order to ensure clear whey by efficiently eradicating residual lipid compounds and microorganisms and facilitating a high-grade WPI followed by several levels

APPLICATIONS OF MEMBRANE TECHNOLOGY

of UF. The protein retentates acquired from UF membrane processing are further processed and subjected to spray-drying in order to develop WPC and WPI based ingredients (Kelly, 2019).

7.4.1 Development of Whey Protein Concentration and Isolates

The ultrafiltration membrane system plays a significant role in the production of whey protein concentrate powder by retaining whey protein in retentates and eventually subjecting it to drying to get the WPC powder. The native WPC comprises a very fine quality amino acid profile with excessive amounts of accessible cysteine and lysine. They are categorized by their protein percentage in dry matter, which varies from 35 to 85%, which are given in Table 7.3. In the manufacture of WPC35 (WPC containing 35% protein content) powder, the whey serum is subjected to an UF of a six-fold concentration to get around 9% of the total dry solid content; however, to produce WPC85 (85% protein content) liquid whey undergoes direct UF, initially becoming concentrated by 20–30 fold, to obtain about 25% solid content (Bylund, 2003). The UF membrane processing of whey is conducted at a 300 kPa inlet pressure with a temperature of < 55°C and a pore size of 250 nm (Wagner, 2001). During UF processing whey retentate contains protein and fat, whereas a large quantity of fluid containing soluble minerals and lactose cross-flow as permeate. Subsequently, the second filtration performed by UF proceeds with the addition of water in the feed (retentate); this technique is known as diafiltration. This is found to be an essential step for diafiltering for the removal of lactose and salts; hence, improve the concentration of the protein in the total dry matter (Bylund, 2003). Microfiltration (MF) also has a significant role in the entire process of the development of WPC85 and WPI (protein content > 90%), as

Table 7.3 Proximate Composition of Various Whey Proteins Concentrates (WPC) and Whey Protein Isolates (WPI) (Bylund, 2003; Kelly, 2019)

Composition	WPC35	WPC65	WPC80	WPI
%Protein dry basis	35	65	80	>90
Fat	2.2	5.5	5.6	1
Ash	7.8	4	3.9	4
Lactose	46.6	21.2	3.3	1
Moisture	4.6	4.1	4.2	3

it facilitates the retention of fat molecules and bacterial cells in retentate form and allows them to pass the permeate containing a high concentration of whey protein. Whey subjected to MF promotes fat reduction from > 7% to < 0.4% (Bylund, 2003). Afterward, the defatted MF permeates are channeled toward the second UF concentration wherein the diafiltration technique is implemented. However, the mixture of fat residues and microbial cells in the MF retentate is further collected and treated as a by-product obtained through WPI processing, which contains a high amount of fat called pro cream. A high amount of denatured whey protein aggregate cannot cross the membrane, therefore, the MF membrane retains the denatured whey protein along with fat in retentate form. However, the denaturation of whey protein can easily be minimized by pasteurization to improve the yield of WPI (Wang and Guo, 2019).

7.4.2 Recovery of Whey Protein Fractionations

Whey protein concentrates (WPCs)/isolates (WIPs) are now commercially viable for the consumer in a dried substance, which varies from 35 to 85% and > 90%, respectively, as shown in Table 7.3. Whey contains major protein types: α-lactalbumin (α-LA), β-lactoglobulin (β-LG), bovine serum albumin (BSA) and chain immunoglobulins (Igs); lactoferrin (LF) and lactoperoxidase proteins are also present in minor quantities, shown in Table 7.1 (Bonnaillie and Tomasula, 2008). Whey may also contain compounds such as glycomacropeptides (GMPs) and other low-molecular-weight components developed during the cheese manufacturing process by the enzymatic degradation of caseins (De Wit, 1989).

Recently the UF membrane technique has been widely applied for whey protein fractionation, which depends on the size variations among the types of proteins that exist in whey because the molecular masses of the principal whey proteins, i.e., α-lactalbumin (α-LA) is 14.4 kDa and β-lactoglobulin (β-LG) is 18.4 kDa. Various investigations concluded that β-LG possesses a molecular weight cutoff (MWCO) of 30–50 kDa, which is majorly restricted by the UF membrane; whereas, α-LA is easily permeated as compared to β-LG. Various researches have been conducted in order to fractionate α-LA and β-LG by using a 30 kDa UF. Cheang and Zydney (2004) have described the process of retaining the β-LG rich fraction, whereas α-LA went with permeate by applying a two-stage tangential flow technique using a 30 kDa UF and a 100 kDa UF applied to refine BSA. However, Marella et al. (2011) investigated the effects of an UF membrane using a molecular weight cutoff (MWCO) of 50 kDa, 100 kDa and 300 kDa

APPLICATIONS OF MEMBRANE TECHNOLOGY

for the separation of α-LA from whey obtained from the processing of cheddar cheese and the yield of pure α-LA acquired was 63% via 50 kDa UF membrane, and hence they concluded that the purity of α-LA would be lower when using looser UF membranes (Etzel and Arunkumar, 2015). The efficient fractionation of whey protein needs proper regulation of the filtration velocity as well as buffer concentration at each UF level. Mehra and Kelly (2004) reported that the utilization of SCT Ceraver ceramic membrane of 0.1 µm combined with Tetra Pak Alcross MFS-19 at a constant TMP control helped in achieving a better separation of high-molecular-weight components, i.e., BSA and Igs from whey with a pH of 5.0. Recently, the combination of membrane pore size with the exploitation of whey protein electrostatic properties via pH-mediation has demonstrated a key potential for influencing the hydrodynamic radius of each protein fraction for efficient isolation through MF and DF; hence, this shows the huge difference between the molecular size of Igs and the 0.1 µm pore size of MF.

Whereas, during the subsequent stage of separation, the above-mentioned MF whey permeates through the UF membrane consisting of three Koch hollow-fiber membranes with MWCO of 30, 50, and 100 kDa, to separate α-La and β-Lg competently with an elimination value of 0.78 and 0.98, respectively. It has been observed that the yield of α-La is higher when applying a 30 kDa membrane, in contrast to lesser retention of α-La on bigger MWCO UF membranes; whereas, a huge simultaneous permeation of β-Lg rebuffs the possibilities of higher α-La enrichment (Kelly, 2019).

Many researchers concluded the need for the diafiltration technique throughout whey protein fractionation from a tight membrane with the combination of other approaches such as the chromatographic technique to intensify the pureness of distinct proteins in whey. Due to these constricted membranes, its inherent protein possesses a lower rate of transmission, which is why a huge amount of water is required to pass protein fractions through these confined membrane structures. Whereas, if the membrane has selectivity for retaining a particular protein and for other protein molecules to permeate completely there will be no need to perform diafiltration, and the separation can be performed efficiently (Etzel and Arunkumar, 2015). Dual-stage UF processing using a flat-disk membrane of 30 and 10 kDa in an agitated rotating disk module suggested that the rotation of the membrane significantly improved flux and was extremely effective in dropping concentration polarization. During the initial state of UF processing whey was exposed to a 30 kDa membrane, where most components, such as lactoferrin (Lf), Igs

APPLICATIONS OF MEMBRANE TECHNOLOGY IN WHEY PROCESSING

and BSA were turned down; wherein, the successive stage of 10 kDa UF processing rejected α-La and β-Lg to the permeate to produce a permeate with much less concentration of protein. Higher fluxes were determined at a ≥ 300 rpm membrane rotation and feed of pH 2.8 to confine TMP to a range of 390–490 kPa (4–5 kg/cm²). During 10 kDa membrane processing, i.e., a final stage 75% concentration of β-Lg (on total protein basis) achieved at TMP of 390 kPa (4 kg/cm²) in an agitated rotating-disk module together with a 600 rpm membrane rotation speed (Bhattacharjee et al., 2006). In contrast, the filtrate obtained through ion-exchange membrane chromatography contained 87.6% pure β-Lg during the separation of a protein component from WPC by application of the aforementioned dual-stage UF combined with an agitated rotating disk accompanied by ion-exchange membrane chromatography equipped with Vivapure Q Mini-H column (Bhattacharjee et al., 2006). Xu et al. (2000) demonstrated the method of anion-exchange chromatography to isolate α-La, β-Lg and BSA and applied the UF technique in order to purify Igs. Saufi and Fee (2013) found that methodologies such as cross-flow cation-exchange chromatography can be utilized to fractionate lactoferrin from whey. Several researchers stated that nowadays ion-exchange chromatography is the most common technique used for the separation of GMP (Doultani et al., 2003, 2004). The chromatographic process combined with membrane separation for the fractionation of LP and LF is simply based on the principle of an isoelectric point at a pH of 9.0–9.5 (alkaline medium). Separation of LF and LP performed via cation-exchange resins, which are especially designed for adsorption of particular protein fraction; wherein, the molecules of LP and LF proteins bind to cation exchanger functional groups carrying a negative charge. Hence, it will lead to the fixing of these molecules on the ion-exchange resin by the charge interaction force, whereas the other protein fraction carrying negative charge will permeate. A cross-flow MF membrane with a 1.4 mm pore size applied under a uniform transmembrane pressure is considered an efficient method for obtaining whey free from particles in about 16 hours with a stable flux of 1200–1500L/m²h; it also prevents the ion-exchange column from the increment of back pressure. As compared to native whey the ion-exchange process is more capable of concentrating LP and LF by a factor of around 500, and in addition to this the UF process with diafiltration increases the yield of the pure protein fractions by up to 95%; thereafter cross-flow microfilters are used with pore sizes of 0.1–0.2 mm to achieve a sterile filtration; hence, the concentrated fraction of the protein is spray-dried (Bylund, 2003).

7.5 CONCENTRATION AND PURIFICATION OF LACTOSE

Whey contains a high amount of lactose which is utilized as a significant ingredient in various pharmaceutical and food products development, so, therefore, the recovery of lactose is essential due to its importance, and it also minimizes the BOD of wastewater and, hence, reduces pollution issues (Souza et al., 2010). The combination of various applied membranes, such as MF, UF, IE and RO, followed by spray-drying can recover and purify lactose from whey (shown in Figure 7.3); wherein, MF and UF are used for fat and protein separation, IE is used for the desalination of whey and RO is applied for concentration and purification of the lactose (Souza et al., 2010). Whey is first subjected to a centrifugal force to eliminate suspended solid particles followed by a UF hollow-fiber membrane system with MWCO 500 kDa, where the UF membrane is made of polyethersulfone, and the system operated at a temperature of 25°C and pressure of 1.5 bar. The lactose reduction is around 0.145; thus, the lower purity of the lactose obtained at a concentration near 88.5%. Afterward, the first UF permeate is sent to a second UF process combined with the diafiltration technique. Many pieces of research have been conducted to compare the two kinds of UF systems, where first is made of a spiral-wound polyethersulfone membrane of 5–8 kDa MWCO and the second is made up of a cassette polyethersulhone membrane with 5 kDa MWCO. The research infers that the UF membrane of cassette polyersulhone with a smaller MWCO, i.e., 5 kDa, showed improved retention of protein and reduced the rejection of lactose. The permeate obtained from the second UF system is subjected to IE for further purification and to remove salts

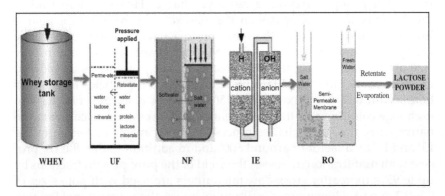

Figure 7.3 Application of membrane technology in lactose recovery.

from the whey followed by RO and spray-drying. A processing system consisting of combined membrane processes, such as MF, UF, IE followed by RO, is considered as the most efficient method for lactose recovery and purification in which MF (titanium oxide tubular membrane) has a 0.2 μm nominal pore diameter and UF has a MWCO of 5 kDa. Hence, the overall recovery of lactose is between approximately 74 and 99.8% pure lactose acquired by this technique (Pruksasri, 2015).

7.6 WHEY DEMINERALIZATION (DESALINATION)

Whey contains a good amount of mineral content, where approximately 10–12% mineral content on a dry basis has been measured, as shown in Table 7.4. The fairly high mineral content present in whey limits its utilization as a component in drugs and food products. Many applications have been found for partially and demineralized whey powders or concentrates, such as in the manufacturing of bakery products, ice-creams and the development of infant food products, respectively. Demineralization is a process applied for the elimination of mineral substances from whey along with amounts of organic ions, such as citrates and lactates. Removal of mineral salt from whey to around 25–30% is termed as partially

Table 7.4 Comparison in the Composition of Mineralized and Various Demineralized Whey Proteins in Market

Composition (Weight %)	Mineralized WP	Demineralized WP		
		NF	ED	IE
Demineralized %	–	<35%	70%	>90%
Moisture content	3	3	3	3
Lactose	73	74	77	>80
Fat	1	1	1	1
Total protein	13	16	14.5	13
Ash content	8	6	4.5	<1
Magnesium	0.6	0.02	0.008	0.007
Calcium	0.2	0.08	0.04	0.02
Phosphorus	0.6	0.5	0.03	0.01

(Bylund, 2003; Wit, 2001; Šímová et al., 2010)

APPLICATIONS OF MEMBRANE TECHNOLOGY

demineralized whey powder or concentrate, which can be further used in the production of various food products such as bakery items, ice-creams, etc. Whereas, more than a 90% reduction of mineral salt from whey is known as highly demineralized whey powder or concentrate, which is utilized as an ingredient in infant formula and other products (Bylund, 2003). Cross-flow pressure-driven membrane systems such as NF (nanofiltration) – an intermediate of RO and UF – is widely applied for the partial demineralization of whey fluid, and it can also be used as a substitute technique for more traditional methods of water treatment due to its high energy efficiency and the simplicity of the process for softening portable water for industrial production (Alkhatim et al., 1998). The significance of NF is that the membrane is extremely efficient toward the eradication of organic components present in whey with molecular weights ranging from 300–1000 Da, such as protein, lactose, etc. (Horst et al., 1995). A NF membrane process is selectively permeable for monovalent ions and their salts (like NaCl, KCl, etc.); however, it's capable of retaining divalent ions such as calcium, magnesium and zinc, which are more nutritionally relevant to human health (Rice et al., 2005 and Van der et al., 2004). A NF membrane is commonly designed in a spiral manner with polyamide, wherein a 200 nm thickness interfacial polyamide layer is formed. Whereas, a more porous interlayer of the membrane is composed of polysulfone and nonwoven polyester, which acts as a supporting and backing material, respectively. The feed pressure during NF processing ranges between 5–20 bar, and operation is usually carried out in a batch mode, wherein the feed is subjected to continuous recycling until the desired amount of salt is eliminated and/or sufficient amount of lactose has been concentrated. This process is followed by diafiltration to increase the efficiency of the process for the continuous permeation of salts, such as NaCl, etc., from the NF membrane (Kentish and Rice, 2015). The recovery of mineral salts as a NF retentate in the form of precipitates can be achieved by adjusting the pH to 7–7.2 via addition of pyrophosphate or tetrasodium pyrophosphate (TSPP), which is then incubated at 60–80°C (Vembu and Rathinam, 1997). Though, by subjecting whey fluid to a NF membrane, only a 30% demineralization is attained by reducing about 70% of the monovalent ions and their salt. To achieve a higher degree of the demineralization of whey, several techniques have been utilized by combining NF with RO, electrodialysis (ED) and ion-exchange (IE) (Greiter et al., 2002). ED and IE are the most convenient and extensively used techniques to acquire a higher degree of demineralized whey concentrate or powder (Šímová et al., 2010).

When liquid whey is subjected to an ED treatment, the tiny electrically charged mineral ions pass through a semi-permeable membrane under the influence of the driving force, i.e., direct current (DC) and electric potential (Bylund, 2003) whereas, bigger molecules restricted by the membrane are left behind in the dilute stream. The retained large particles of organic compounds settle on the surface of the membrane, which may reduce the process efficiency by blocking the path of the mineral ions across the membrane; hence, leading to fouling (Šímová et al., 2010). The scaling layer of these salts Ca^{2+}, Mg^{2+}, SO_4^{2-}, PO_4^{3-} or CO_3^{2-} has been found during ED of whey (Bribiesca et al., 2006; Ruiz et al., 2007; Shee et al., 2008). Selection of suitable processing conditions such as spacer thickness, flow rate, type of spacer, etc. are necessary to minimize membrane fouling issues (Šímová et al., 2010).

The ED processing unit contains a number of partitions divided by the alternative placement of anion and cation-exchange membranes, which are placed approximately < 1 mm from each other, with the electrodes placed at the end, as many as cell pairs can be put in between electrodes as needed (Bylund, 2003). Under the influence of direct current, anions and cations try to migrate from anode and cathode, respectively, as shown in Figure 7.4.

Though, free migration of the anion and cation has not yet been achieved due to the hindrance of the likely charged membrane. Where a positively charged membrane allows cations to pass through, anions get trapped; simultaneously, negatively charged membranes allow the movement of anions but not cations (Šímová et al., 2010). Hence demineralization of whey is attained based on the amount of ash present, flow viscosity, residence time and current density. The ED of whey can be done either in a continuous or in a batch system, but the batch system is more efficient for the reduction of the mineral content of whey to more than 70% because it continuously circulates until the desired ash level is achieved.

The retention time of the batch system is around 5–6 hours for up to a 90% demineralization of whey at a temperature of 30–40°C. Before whey is subjected to an ED unit, the clarification and pre-concentration (20–30%) of the whey is necessary for capacity utilization and to minimize power consumption (Bylund, 2003). ED is one of the most efficient techniques for whey demineralization up to 70%, but in contrast IE is known for a higher mineral reduction in whey of more than 90%, as shown in Figure 7.4. In this process, a fixed-bed separation technology containing exchange resins is used (Greiter et al., 2002). Exchange resins have a strong potency to

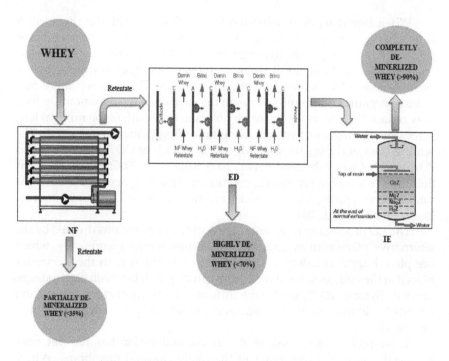

Figure 7.4 A systematic approach to the manufacturing of various demineralized whey products in the market.

bind ions on their surface from an aqueous medium. The anion-exchange resins trap negatively charged ions and cation-exchange resins attach to positively charged ions. The capacity of ion-exchange resins is limited, and after their complete saturation, regenerated resins are added before reuse (Greiter et al., 2002).

7.7 CHALLENGES RELATED TO MEMBRANE PROCESSING

Two main issues develop during membrane processing, known as concentration polarization and membrane fouling, which are considered inevitable occurrences that are related to all types of membrane processing, but they are especially observed when milk or whey is subjected to UF and MF processing (Steinhauer et al., 2015). Concentration polarization and membrane fouling are both completely different from each other, but results are similar as both phenomena affect process efficiency. The concentration polarization is totally reversible in nature, and it occurs due to mass transfer constraints in the module, but when completely developed it does not reduce flux with time. Whereas, when membrane fouling occurs, it decreases flux over time even though all the processing parameters remain constant and there is also the requirement of certain types of chemical or mechanical treatments to clean the membrane for further processing because it is an irreversible phenomenon (Sablani et al., 2001; Saxena, et al., 2009). Fouling of the membrane refers to blockage of the pores of the membrane due to the deposition of solute particles present in the feed stream, such as protein (James and Chen, 2003), mineral (Popovic et al., 2010), bacterial biofilm (Anand et al., 2012), etc., which leads to cake formation on the surface of the membrane and depth fouling (Tang et al., 2009).

7.7.1 Concentration Polarization

When feed is subjected to membrane processing, the partial or complete retention of solute components in the feed move closer to the surface of the membrane and gets accumulated on it. Concentration polarization exerts extra hydraulic resistance to the flow stream as well as developing osmotic pressure, which eventually decreases the effectiveness of transmembrane pressure during the membrane filtration process (Kanani, 2015). The developed layer of concentration polarization has a minimal to moderate effect on the pressure-controlled region, which is generally categorized in either low feed concentration, high cross-flow rate or low pressure. The concentration polarization layer is the accumulation of solute particles that develops a concentration gradient in the hydrodynamic boundary layer. The movement of solute particles through the boundary layer is done by convection; however, their reverse flow against the boundary layer in the direction of the bulk feed happens due to the diffusion process. The boundary layer possesses quietly viscous and gelatinous characteristics, which exert an additional barrier to the filtrate flow flux with time (Kanani, 2015).

7.7.2 Membrane Fouling

Membrane fouling is considered as the major obstacle to achieving efficient membrane processing, and it occurs due to the adsorption and deposition of the feed stream particles on the surface or inside of the pore of the membrane; hence, the clogging of the membrane will lead to an excess of hydraulic resistance with a reduction in permeate flux over time; while the other processing parameters remain constant (such as temperature, flow rate, feed concentration and pressure) (Kanani, 2015). Hence fouling of the membrane eventually affects process efficiency and its productivity, reduces membrane lifetime (because of the treatment with harsh agents during cleaning), and hampers the filtration ability of the membrane. Major microfiltration and ultrafiltration membrane processes encounter membrane fouling (Huimin et al., 2001). There are different fouling mechanisms which can be broadly classified under two groups:

1. External or dynamic membrane fouling (also known as particulate deposition and filtration-induced macrosolute) is often observed in a non-flowing membrane system where continuous deposition of feed particles on the surface of the membrane leads to the formation of a cake layer (Huimin et al., 2001; Mohammad et al., 2012).
2. Internal or static membrane fouling (also known as particle adsorption and macrosolute), refers to the specific intermolecular interaction (both physical and chemical interaction) between the membrane and the feed particles and takes place in the absence of the permeate flux and cause the accumulation of particles in the pores of the membrane resulting in partially and complete pore blockage (Huimin et al., 2001; Mohammad et al., 2012).

Hence, both fouling situations end up with the formation of a gel-like layer near the membrane surface based on the concentration polarization. The evaluation of the fouling of the membrane can be accomplished via regulating the permeate flux by keeping the temperature and pressure constant (Kanani, 2015). The flux and total resistance are inversely proportional to each other; as the flux decreases, the fouling of membrane increases over time. An immediate reduction in flux indicates that the membrane is extremely susceptible to fouling with the specific feed particulates. Whereas, several procedures are there to evaluate the fouling trends of the membrane, such as the plugging index or silt density index, wherein the feed stream are observed for fouling rather than the membrane characteristics (Kanani, 2015).

There are a variety of components present in the feed stream which interact differently with the membrane according to the operating conditions; thus, it is quite challenging to develop universal theories or laws which appropriately comment on the nature of a membrane and its interaction with feed solutes that are responsible for fouling the membrane. However, some common factors which influence membrane fouling are classified under three groups.

(1) Feed characteristics (such as physicochemical properties of the feed solute particles, salt, concentration and pH). (2) Characteristics of the membrane (such as membrane surface charge, pore size distribution, porosity, pore size and hydrophobicity). (3) Operating parameters (such as cross-flow velocity, transmembrane pressure and temperature) temperature (Cheryan, 1998; Marshall et al., 1993)

Membrane fouling can be overcome by conducting a systematic and consistent cleaning program, such as crossflushing, backpulsing and backwashing (Huimin et al., 2001; James et al., 2003; Mohammad et al., 2012) on the membrane at appropriate time gaps and by using a membrane that is less prone to fouling (Saboya and Maubois, 2000); this can be done by using membrane modules with appropriate channel heights, by a high-pressure application during filtration process (Wakeman and Williams, 2002), application of electric potential, microturbulence, ultrasoundwaves and uniform transmembrane pressure (Duriyabunleng et al., 2001), and by installing ceramic membranes with vibrating and rotating-disc modules (Ding et al., 2002).

7.8 FUTURE TRENDS

The development of innovative techniques and their applications in food processing is a pressing world issue. Presently, dairy processes involve membrane processing, which usually has a low capacity due to the continuous flux reduction with time through the development of membrane fouling; the processes are also high energy-consuming because of the demand to upsurge the cross-flow velocity, which is essential to limiting fouling of the membranes. Moreover, procedures to control fouling have enlarged the intricacies in apparatuses and operations. Apart from fouling, membrane selectivity is the main issue in achieving a desirable end product, such as the concentration and fractionation of specific components from milk and whey. Hence, for the production of economical, readily accessible, high quality and durable membranes, innovative techniques are required. The needs of the end-users and retailers for the availability of fresh products of superior quality and longer shelf life can be better fulfilled by utilizing membrane technology as compared to conventional methods.

7.9 CONCLUSION

Membrane technology is recognized as a revolution in dairy and food processing, where it possesses several benefits, such as being environmentally friendly, waste management, waste treatment, low energy consumption during processing, better product quality with high yield and it is also considered as a "green technology." Thus, these are the main reasons behind the rapid utilization of membrane technology in dairy processing. It also contributes to the innovation of various new products and enhances nutritive value, sensory attributes and yield of existing market products. Since this chapter mainly focused on whey processing using membrane technology, all the membrane processing techniques such as MF, UF, NF, RO and ED and their significance were discussed in respect of various whey products and their development. Though, there are still some factors such as membrane fouling that affect membrane processing; hence, it reduces process efficiency and prevents its further application. The consistent efforts in the creation of membranes with superior characteristics may further enhance their role in dairy processing.

REFERENCES

Abidin, N.S.Z., Hussain, S.A. and Kamal, S.M.M. 2014. Removal of lactose from highly Goat's milk concentration through ultrafiltration membrane. In: *Process and Advanced Materials Engineering, Applied Mechanics and Materials,* edited by I. Ahmed, vol. 625, 596–599. Scientific.net. doi:10.4028/www.scientific.net/AMM.625.596.

Al-Akoum, O., Ding, L., Chotard-Ghodsnia, R., Jaffrin, M.Y. and Gésan-Guiziou, G. 2002. Casein micelles separation from skimmed milk using a VSEP dynamic filtration module. *Desalination*, 144(1–3):325–330.

Alkhatim, H.S., Alcaina, M.I., Soriano, E., Iborra, M.I., Lora, J. and Arnal, J. 1998. Treatment of whey effluents from dairy industries by nanofiltration membranes. *Desalination*, 119(1–3):177–184.

Anand, A., Hassan, A. and Avadhanula, M. 2012. The effects of biofilms formed on whey reverse osmosis membranes on the microbial quality of the concentrated product. *Int J Dairy Technol*, 65(3):451–455.

Andrade, L.H., Mendes, F.D.S., Espindola, J.C. and Amaral, M.C.S. 2014. Nanofiltration as tertiary treatment for the reuse of dairy wastewater treated by membrane bioreactor. *Sep Purif Technol*, 126:21–29.

Andrés, L.J., Riera, F.A. and Alvarez, R. 1995. Skimmed milk demineralization by electrodialysis: Conventional versus selective membranes. *J Food Eng*, 26(1):57–66.

Atra, R., Vatai, G., Bekassy-Molnar, E. and Balint, A. 2005. Investigation of ultra- and nanofiltration for utilization of whey protein and lactose. *J Food Eng*, 67(3):325–332.

Balannec, B., Vourch, M., Rabiller-Baudry, M. and Chaufer, B. 2005. Comparative study of different nanofiltration and reverse osmosis membranes for dairy effluent treatment by dead-end filtration. *Sep Purif Technol*, 42(2):195–200.

Barba, D., Beolchini, F. and Vegliò, F. 2000. Minimizing water use in diafiltration of whey protein concentrates. *Sep Sci Technol*, 35(7):951–965.

Barba, D., Beolchini, F., Cifoni, D. and Veglio, F. 2001. Whey protein concentrate production in a pilot scale two-stage diafiltration process. *Sep Sci Technol*, 36(4):587–603.

Barba, D., Beolchini, F., Cifoni, D. and Veglio, F. 2002. Whey ultrafiltration in a tubular membrane: Effect of selected operating parameters. *Sep Sci Technol*, 37(8):1771–1788.

Bazinet, L., Ippersiel, D., Gendron, C., René-Paradis, J., Tétrault, C. and, Beaudry, J. 2001. Bipolar membrane electroacidification of demineralized skim milk. *J Agric Food Chem*, 49(6):2812–2818.

Bhattacharjee, S., Bhattacharjee, C. and Datta, S. 2006. Studies on the fractionation of β-lactoglobulin from casein whey using ultrafiltration and ion-exchange membrane chromatography. *J Membr Sci*, 275(1–2):141–150.

Bhattacharjee, S., Ghosh, S., Datta, S. and Bhattacharjee, C. 2006. Studies on ultra-filtration of casein whey using a rotating disk module: Effects of pH and membrane disk rotation. *Desalination*, 195(1–3):95–108.

Bonnaillie, Laetitia M. and Tomasula, Peggy M. 2008. Whey protein fractionation. In: *Whey Processing, Functionality and Health Benefits*, edited by Charles I. Onwulata and Peter J. Huth, 15–38. Lowa: John Wiley & Sons.

Bylund, Gösta. 2003. *Dairy Processing Handbook*. Sweden: Tetra Pak Processing Systems AB.

Chai, M., Ye, Y. and Chen, V. 2017. Separation and concentration of milk proteins with a submerged membrane vibrational system. *J Membr Sci*, 524:305–314.

Cheang, B. and Zydney, A.L. 2004. A two-stage ultrafiltration process for fraction-ation of whey protein isolate. *J Membr Sci*, 231(1–2):159–167.

Chen, G.Q., Eschbach, F.I.I., Weeks, M., Gras, S.L. and Kentish, S.E. 2016. Removal of lactic acid from acid whey using electrodialysis. *Sep Purif Technol*, 158:230–237.

Chen, Q., Zhao, L., Yao, L., Chen, Q., Ahmad, W., Li, Y. and Qin, Z. 2018. The application of membrane separation technology in the dairy industry. In: *Technological Approaches for Novel Applications in Dairy Processing*, edited by Nurcan Koca, 23–34. Licensee InTech. doi:10.5772/intechopen.76320.

Cheryan, M. 1998. *Ultrafiltration and Microfiltration Handbook*. Lancaster: Technomic.

Chollangi, A. and Hossain, M.M. 2007. Separation of proteins and lactose from dairy wastewater. *Chem Eng Process*, 46(5):398–404.

Crowley, S.V., Caldeo, V., McCarthy, N.A., Fenelon, M.A., Kelly, A.L. and O'Mahony, J.A. 2015. Processing and protein-fractionation characteristics of different polymeric membranes during filtration of skim milk at refrigeration temperatures. *Int Dairy J*, 48:23–30.

Cuartas-Uribe, B., Alcaina-Miranda, M.I., Soriano-Costa, E., Mendoza-Roca, J.A., Iborra-Clar, M.I. and Lora-Garcia, J. 2009. A study of the separation of lactose from whey ultrafiltration permeate using nanofiltration. *Desalination*, 241(1–3):244–255.

De Wit, J.N. 1989. Functional properties of whey proteins. In: *Developments in Dairy Chemistry-4*, edited by P.F. Fox. New York: Elsevier Applied Science.

Ding, L., Al-Akoum, O., Abraham, A. and Jaffrin, M.Y. 2002. Milk protein concentration by ultrafiltration with rotating disk modules. *Desalination*, 144(1–3):307–311.

Doultani, S., Turhan, K.N. and Etzel, M.R. 2003. Whey protein isolate and glycomacropeptide recovery from whey using ion-exchange chromatography. *J Food Sci*, 68:1389–1395.

Doultani, S., Turhan, K.N. and Etzel, M.R. 2004. Fractionation of proteins from whey using cation exchange chromatography. *Process Biochem*, 39(11):1737–1743.

Duriyabunleng, H., Petmunee, J. and Muangnapoh, C. 2001. Effects of the ultrasonic waves on microfiltration in plate and frame module. *J Chem Eng Japan / JCEJ*, 34(8):985–989.

Espina, V.S., Jaffrin, M.Y. and Ding, L.H. 2009. Comparison of rotating ceramic membranes and polymeric membranes in fractionation of milk proteins by microfiltration. *Desalination*, 245(1–3):714–722.

Etzel, Mark R. and Arunkumar, Abhiram. 2015. Dairy protein fractionation and concentration using charged ultrafiltration membranes. In: *Processing for Dairy Ingredient Separation*, edited by Kang Hu and James M. Dickson, 86–111. Chichester: John Wiley & Sons.

Fauquant, J., Beaucher, E., Sinet, C., Robert, B. and Lopez, C. 2014. Combination of homogenization and cross-flow microfiltration to remove micro-organisms from industrial buttermilks with an efficient permeation of proteins and lipids. *Innov Food Sci Emerg Technol*, 21:131–141.

Fidaleo, M. and Moresi, M. 2006. Electrodialysis applications in the food industry. *Adv Food Nutr Res*, 51:265–360.

Fritsch, J. and Moraru, C.I. 2008. Development and optimization of a carbon dioxide-aided cold microfiltration process for the physical removal of microorganisms and somatic cells from skim milk. *J Dairy Sci*, 91(10):3744–3760.

Gesan-Guiziou, G., Alvarez, N., Jacob, D. and Daufin, G. 2007. Cleaning-in-place coupled with membrane regeneration for reusing caustic soda solutions. *Sep Purif Technol*, 54(3):329–339.

Greiter, M., Novalin, S., Wendland, M., Kulbe, K.D. and Fischer, J. 2002. Desalination of whey by electrodialysis and ion exchange resins: Analysis of both processes with regard to sustainability by calculating their cumulative energy demand. *J Membr Sci*, 210(1):91–102.

Heino, A.T., Uusi-Rauva, J.O., Rantamaki, P.R. and Tossavainen, O. 2007. Functional properties of native and cheese whey protein concentrate powders. *Int J Dairy Technol*, 60(4):277–285.

Hickey, R.M. 2012. The role of oligosaccharides from human milk and other sources in prevention of pathogen adhesion. *Int Dairy J*, 22(2):141–146.

Hinkova, A., Zidova, P., Pour, V., Bubnik, Z., Henke, S., Salova, V. and Kadlec, P. 2012. Potential of membrane separation processes in cheese whey fractionation and separation. *Procedia Eng*, 42:1425–1436.

Hinrichs, J. and Atamer, Z. 2011. Heat treatment of milk sterilization of milk and other products A2 – Fuquay. In: *Encyclopedia of Dairy Sciences*. 2nd ed., edited by W. John, 714–724. San Diego: Academic Press.

Hoffmann, W., Kiesner, C., Clawin-RÄDecker, I., Martin, D., Einhoff, K. and, Lorenzen, P.C. 2006. Processing of extended shelf life milk using microfiltration. *Int J Dairy Technol*, 59(4):229–235.

Huimin, M., Hakimb,L.F., Bowmana, C.N. and Davis, R.H. 2001. Factors affecting membrane fouling reduction by surface modification and backpulsing. *J Membr Sci*, 189:255–270.

Hurt, E., Zulewska, J., Newbold, M. and Barbano, D.M. 2010. Micellar casein concentrate production with a 3X, 3-stage, uniform transmembrane pressure ceramic membrane process at 50 degrees C. *J Dairy Sci*, 93(12): 5588–5600.

James, B.J. and Dong Chen, X. 2003. Membrane fouling during filtration of milk-a microstructural study. *J Food Eng*, 60(4):431–437.

James, B.J., Jing, Y. and Chen, X.D. 2003. Membrane fouling during filtration of milk—A microstructural study. *J Food Eng*, 60(4):431–437.

Jevons, M. 1997. Separate and concentrate. *Dairy Ind Int*, 8:19–21.

Kanani, Dharmesh. 2015. Membrane fouling: A challenge during dairy ultrafiltration. In: *Processing for Dairy Ingredient Separation*, edited by Kang Hu and James M. Dickson, 65–85. Chichester: John Wiley & Sons.

Kentish, Sandra E. and Rice, G. 2015. Demineralization of dairy streams and dairy mineral recovery using nanofiltration. In: *Processing for Dairy Ingredient Separation*, edited by Kang Hu and James M. Dickson, 112–138. Chichester: John Wiley & Sons.

Kiesner, C. and Eggers, R. 1994. Concept of a sterile concentration process for milk by multistage flash evaporation. *Chem Eng Technol*, 17(6):374–381.

Kolfschoten, R.C., Janssen, A.E.M. and Boom, R.M. 2011. Mass diffusion-based separation of sugars in a microfluidic contactor with nanofiltration membranes. *J Sep Sci*, 34(11):1338–1346.

Lawrence, N.D., Kentish, S.E., O'Connor, A.J., Barber, A.R. and Stevens, G.W. 2008. Microfiltration of skim milk using polymeric membranes for casein concentrate manufacture. *Sep Purif Technol*, 60(3):237–244.

Lewis, M.J. 1990. *Physical Properties of Foods and Food Processing System*. Chichester, UK: Ellis Horwood.

Lipnizki, F., Hausmanns, S., Ten, P.K., Field, R.W. and Laufenberg, G. 1999. Organophilic pervaporation: Prospects and performance. *Chem Eng J*, 73(2):113–129.

APPLICATIONS OF MEMBRANE TECHNOLOGY

Liu, D., Li, J., Zhang, J., Liu, X., Wang, M. and, Hemar, Y. 2017. Effect of partial acidification on the ultrafiltration and diafiltration of skim milk: Physicochemical properties of the resulting milk protein concentrates. *J Food Eng,* 212:55–64.

Lorenzen, P.C., Clawin-Raedecker, I., Einhoff, K., Hammer, P., Hartmann, R. and, Hoffmann, W. 2011. A survey of the quality of extended shelf life (ESL) milk in relation to HTST and UHT milk. *Int J Dairy Technol,* 64(2):166–178.

Luo, J., Ding, L., Wan, Y., Paullier, P. and Jaffrin, M.Y. 2010. Application of NF-RDM (nanofiltration rotating disk membrane) module under extreme hydraulic conditions for the treatment of dairy wastewater. *Chem Eng J,* 163(3):307–316.

Luo, J., Nordvang, R.T., Morthensen, S.T., Zeuner, B., Meyer, A.S. and, Mikkelsen, J.D. 2014. An integrated membrane system for the biocatalytic production of 3′-sialyllactose from dairy by-products. *Bioresour Technol,* 166:9–16.

Macwan, S.R., Dabhi, B.K., Parmar, S.C. and Aparnathi, K.D. 2016. Whey and its utilization. *Int J Currmicrobiol App Sci,* 5(8):134–155.

Madaeni, S.S., Tavakolian, H.R. and Rahimpour, F. 2011. Cleaning optimization of microfiltration membrane employed for milk sterilization. *Sep Sci Technol,* 46(4):571–580.

Marella, C., Muthukumarappan, K. and Metzger, L.E. 2011. Evaluation of commercially available wide-pore size ultrafiltration membranes for production of alpha-lactalbumin enriched whey protein concentrate. *J Dairy Sci,* 94(3):1165–1175.

Marella, C., Muthukumarappan, K. and Metzger, L.E. 2013. Application of membrane separation technology for developing novel dairy food ingredients. *J Food Process Technol,* 4:269.

Marshall, A.D., Munro, P.A. and Tragardh, G. 1993. The effect of protein fouling in microfiltration and ultrafiltration on permeate flux, protein retention and selectivity: A literature review. *Desalination,* 91(1):65–108.

Marshall, K. 2004. Therapeutic applications of whey protein. *Altern Med Rev,* 9(2):136–156.

Mehra, R. and Kelly, P.M. 2004. Whey protein fractionation using cascade membrane filtration. In: *Proceedings of IDF World Dairy Summit 2003 Symposium 'Advances in Fractionation and Separation: Processes for Novel Dairy Applications',* 40–44. Bulletin of the International Dairy Federation No, 389.

Mohammad, A.W., Ching, Y.N., Lim, Y.P. and Gen, H.N. 2012. Ultrafiltration in food processing industry: Review on application, membrane fouling, and fouling control. *Food Bioprocess Technol,* 5(4):1143–1156.

Mohsenin, N.N. 1980. *Physical Properties of Plant and Animal Materials: Structure, Physical Characteristics and Mechanical Properties.* New York: Gordan and Breach.

Morr, C.V. and Brandon, S.C. 2008. Membrane fractionation processes for removing 90 to 95% of the lactose and sodium from skim milk and for preparing lactose and sodium-reduced skim milk. *J Food Sci,* 73:C639–C647.

Mulder, M. 1991. *Basic Principles of Membrane Technology.* Norwell: Kluwer Academic Publishers.

Namvar-Mahboub, M. and Pakizeh, M. 2012. Experimental study of lactose hydrolysis and separation in CSTR-UF membrane reactor. *Braz J Chem Eng*, 29(3):613–618.

Nissar, N., Hameed, O.B. and Nazir, F. 2018. Application of Membrane Technology in Food processing Industries: A review. *Int J Adv Res Sci Eng*, 7:2106–2114.

Oliveira, D.L., Wilbey, R.A., Grandison, A.S. and Roseiro, L.B. 2015. Milk oligosaccharides: A review. *Int J Dairy Technol*, 68(3):305–321.

Overington, A.R., Wong, M. and Harrison, J.A. 2011. Effect of feed pH and non-volatile dairy components on flavour concentration by pervaporation. *J Food Eng*, 107(1):60–70.

Phil, Kelly. 2019. Manufacture of whey protein products: Concentrates, isolate, whey protein fractions and Microparticulated. In: *Whey Proteins: From Milk to Medicine*, edited by Hilton C. Deeth and Nidhi Bansal, 95–122. London: Andre Gerhard Wolff.

Popovic, S., Djuric, M., Milanovic, S., Tekic, M.N. and Lukic, N. 2010. Application of an ultrasound field in chemical cleaning of ceramic tubular membrane fouled with whey proteins. *J Food Eng*, 101(3):296–302.

Pouliot, Y., Wijers, M.C., Gauthier, S.F. and Nadeau, L. 1999. Fractionation of whey protein hydrolysates using charged UF NF membranes. *J Membr Sci*, 1(1–2):105–114.

Pruksasri, Suwattana. 2015. Dairy stream lactose fractionation/concentration using polymeric ultrafiltration membrane. In: *Processing for Dairy Ingredient Separation*, edited by Kang Hu and James M. Dickson, 35–66. Chichester: John Wiley & Sons.

Rahimi, Z., Zinatizadeh, A.A. and Zinadini, S. 2016. Milk processing wastewater treatment in a bioreactor followed by an antifouling O-carboxymethyl chitosan modified Fe3O4/PVDF ultrafiltration membrane. *J Ind Eng Chem*, 38:103–112.

Rektor, A. and Vatai, G. 2004. Membrane filtration of Mozarella whey. *Desal*, 162(279):286.

Rice, G., Kentish, S., and Vivekanand, V. 2005. Membrane-based dairy separation: A comparison of nanofiltration and electrodialysis. *Dev Chem Eng Min Process*, 13(1–2):43–54.

Richards, N.S.P.S. 2002. Soro Lácteo – Perpectivas Industriais e Proteção aoMeio Ambiente. *Food Ingred*, 17:20–27.

Rinaldoni, A.N., Campderros, M., Menendez, C.J. and Padilla, A.P. 2009. Fractionation of skim milk by an integrated membrane process for yoghurt elaboration and lactose recuperation. *Int J Food Eng*, 5(3):1–17.

Rinn, J.C., Morr,C.V., Seo, A. and Surak, J.G. 1990. Evaluation of nine semi-pilot scale whey pretreatment modifications for producing whey protein concentrate. *J Food Sci*, 55(2):510–515.

APPLICATIONS OF MEMBRANE TECHNOLOGY

Ruiz, B., Sistat, P., Huguet, P., Pourcelly, G., Araya-Farias, M. and Bazinet, L. 2007. Application of relaxation periods during electrodialysis of a casein solution: Impact on anion-exchange membrane fouling. *J Membr Sci*, 287(1):41–50.

Sablani, S.S., Goosen, M.F.A., Al-Belushi, R. and Wilf, M. 2001. Concentration polarization in ultrafiltration and reverse osmosis: A critical review. *Desalination*, 141(3):269–289.

Saboya, L.V. and Maubois, J.L. 2000. Current developments of microfiltration technology in the dairy industry. *Lait*, 80(6):541–553.

Samtlebe, M., Wagner, N., Neve, H., Heller, K.J., Hinrichs, J. and Atamer, Z. 2015. Application of a membrane technology to remove bacteriophages from whey. *Int Dairy J*, 48:38–45.

Saufi, S.M. and Fee, C.J. 2013. Recovery of lactoferrin from whey using crossflow cation-exchange mixed matrix membrane chromatography. *Sep Purif Technol*, 77(1):68–75.

Saxena, A., Tripathi, B.P., Kumar, M. and Shahi, V.K. 2009.Membrane-based techniques for the separation and purification of proteins: An overview. *Adv Colloid Interface*, 145(1–2):1–22.

Schmidt, V.S.J., Kaufmann, V., Kulozik, U., Scherer, S. and Wenning, M. 2012. Microbial biodiversity, quality and shelf life of microfiltered and pasteurized extended shelf life (ESL) milk from Germany, Austria and Switzerland. *Int J Food Microbiol*, 154(1–2):1–9.

Shee, F.L.T., Angers, P. and Bazinet, L. 2008. Microscopic approach for the identification of cationic membrane fouling during Cheddar cheese whey electro-acidification. *J Colloid Interface Sci*, 322(2):551–557.

Šímová, H., Kysela, V. and Černín, A. 2010. Demineralization of natural sweet whey by electrodialysis at pilot-plant scale. *Desalin Water Treat*, 14(1–3):170–173.

Smithers, G.W. 2008. Whey and whey proteins-from 'gutter-to-gold.' *Int Dairy J*, 18(7):695–704.

Solanki, G. and Rizvi, S.S.H. 2001. Physico-chemical properties of skim milk retentates from microfiltration. *J Dairy Sci*, 84(11):2381–2391.

Souzaa, R.R. de, Bergamascoa, R., Costab, S.C. da, Fengc, X., Fariaa, S.H.B. and Gimenesa, M.L. 2010. Recovery and purification of lactose from whey. *Chem Eng Process*, 49(11):1137–1143.

Steinhauer, T., Marx, M., Bogendörfer, K. and Kulozik, U. 2015. Membrane fouling during ultra-and microfiltration of whey and whey proteins at different environmental conditions:the role of aggregated whey proteins as fouling initiators. *J Membr Sci*, 489:20–27.

Tang, X., Fint, S.H., Brooks, J.D. and Bennett, R.J. 2009. Factors affecting the attachment of micro-organisms isolated from ultrafiltration and reverse osmosis membranes in dairy processing plants. *J Appl Microbiol*, 107(2):443–451.

Thien, T.L., Angeli, D.C. and Van, M.B. 2014. Membrane separations in dairy processing. *J Food Res Tech*, 2:1–14.

Tomasula, P.M., Mukhopadhyay, S., Datta, N., Porto-Fett, A., Call, J.E. and, Luchansky, J.B. 2011. Pilot-scale crossflow-microfiltration and pasteurization to remove spores of *Bacillus anthracis* (Sterne) from milk. *J Dairy Sci*, 94(9):4277–4291.

Tunick, Michael H. 2019. Whey protein production and utilization: A brief history. In: *Whey Protein Production, Chemistry,Functionality, and Applications*, edited by Mingruo Guo, 1–14. Chichester: John Wiley & Sons.

Van der Bruggen, B., Koninckx, A. and Vandecasteele, C. 2004. Separation of monovalent and divalent ions from aqueous solution by electrodialysis and nanofiltration. *Water Res*, 38(5):1347–1353.

van der Horst, H.C., Timmer, J.M.K., Robbertsen, T. and Leenders, J. 1995. Use of nanofiltration for concentration and demineralization in the dairy industry: Model for mass transport. *J Membr Sci*, 104(3):205–218.

Vembu, R. and Rathinam,V. 1997. Separation of minerals from whey permeate. *United States Patent*.

Vourch, M., Balannec, B., Chaufer, B. and Dorange, G. 2005. Nanofiltration and reverse osmosis of model process waters from the dairy industry to produce water for reuse. *Desalination*, 172(3):245–256.

Wagner, J. 2001. *Membrane Filtration Handbook: Practical Hints and Tips*. Hopkins, MN: Osmonics.

Wakeman, R.J. and Williams, C.J. 2002. Additional techniques to improve microfiltration. *Sep Purif Technol*, 26(1):3–18.

Wang, Guorong and Guo, Mingruo. 2019. Manufacturing technologies of whey protein products. In: *Whey Protein Production, Chemistry,Functionality, and Applications*, edited by Mingruo Guo, 13–38. Chichester, UK: John Wiley & Sons.

Winston Ho, W.S. and Sirkar, K.K. 1992. *Membrane Handbook*, edited by W.S. Winston Ho and K.K. Sirkar, 3–16. New York, NY: Van Nostrand Reinhold.

Wit, J.N.D.E. 2001. *Lecture's Handbook on Whey and Whey Products*. Belgium: European Whey Product Association.

Xu, Y., Sleigh, R., Hourigan, J. and Johnson, R. 2000. Separation of bovine immunoglobulin G and glycomacropeptide from dairy whey. *Process Biochem*, 36(5):393–399.

Yorgun, M.S., Balcioglu, I.A. and Saygin, O. 2008. Performance comparison of ultrafiltration, nanofiltration and reverse osmosis on whey treatment. *Desalination*, 229(1–3):204–216.

8

Membrane Technology for Degumming, Dewaxing and Decolorization of Crude Oil

M. Selvamuthukumaran

Contents

8.1	Introduction	201
8.2	Membranes Used for the Degumming Process	203
8.3	Experimental Procedure for Oil Degumming Process	204
8.4	Dewaxing and Decolorization of Oil	205
8.5	Conclusions	208
References		208

8.1 INTRODUCTION

Vegetable oil in its crude form is comprised of phospholipids and several impurities like sterols, colors, diglycerides, free fatty acids, etc. (Figure 8.1). The phospholipids are a major component that may vary for different kinds of edible vegetable oils. Soybean oil contains 3–3.5% phospholipid, while groundnut oil contains fewer phospholipids to the extent of 0.2%. During the storage of crude oil, the phospholipids present in vegetable oil undergo precipitation, which creates turbidity and causes

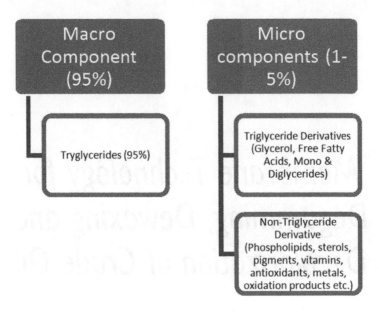

Figure 8.1 Composition of crude oil.

off-flavors, making the stored oil unfit for further consumption. The color gets degraded, and free fatty acids may undergo rancidity making the oil unacceptable and reducing the safety of consumption. Therefore in order to enhance storage stability and quality, crude oil needs a refining process. During the refining process, the various impurities present in oils, *viz.* phosphor lipids, free fatty acids, color and off-flavor components, and plant waxes, can be leached out and reduced; hence the quality of oil can be improved to a greater extent.

In past decades, technologies were limited, and therefore the refining process was carried out by a water and acid degumming process. There are several disadvantages (Table 8.1) associated with this method, such as the oil yield will be less andthe process leads to greater acidic wash water generation; moreover, the desired phospholipid level cannot be successfully reduced by using this conventional process. Currently, the conventional process has been replaced with a membrane technology for more efficient impurities removal (Table 8.2). This technology is convenient, eco-friendly and consumes less energy when compared to traditional conventional processes. Koseoglu et al. (1990) carried out a membrane separation technique at a pilot level for degumming raw vegetable oil and

Table 8.1 Disadvantages of Conventional Methods of Oil Refining

Energy consumption is more

Needs more water usage

Use of chemicals may lead to the retention of some residues in the finished product

More waste is generated

It needs a high temperature for the refining process

Heat labile components were destroyed during processing

Not eco-friendly

Table 8.2 Different Types of Processes Used for Degumming of Crude Vegetable Oil With its Efficiency Comparison

Type of Process	Extent of Impurities Removal	Commercial Feasibility
Conventional Process	Less	Less
Membrane Process	More	More

they concluded that the process could be efficiently adopted to remove the phospholipids, waxes and impurities from crude vegetable oil. This was further proven by several other researchers: Subrahmanyam and Bhowmick (1999) used a ceramic membrane for degumming crude rice bran oil; Subramanian et al. (1999) demonstrated that use of surfactants could enhance the removal of phospholipids during the membrane processing of oil; Kale et al. (1999) removed the free fatty acids from oil by a solvent extraction process followed by passing it through the membrane to remove the FFA to a greater extent.

8.2 MEMBRANES USED FOR THE DEGUMMING PROCESS

There are different types of membranes used for successfully degumming oil (Figure 8.2). The ultrafiltration membrane, i.e., UF I, II, III and nanofiltration membrane (NF), was successfully employed for the degumming of crude edible oil (Desai et al., 2002). The ultrafiltration membrane, i.e., UF I, II, is made up of a polysulfone polymer with a molecular weight cut off (MWCO) range of 250,000 and 1,00,000. The ultrafiltration membrane III is

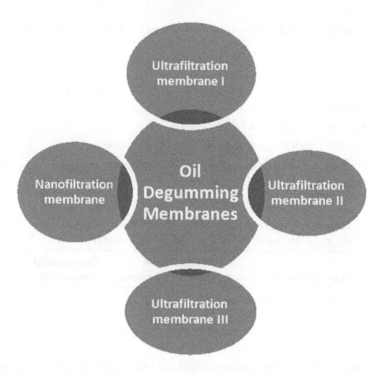

Figure 8.2 Different types of membrane used for degumming of crude vegetable oil.

made up of polyethersulfone with a MWCO of 10,000, and the nanofiltration membrane is made up of cellulose triacetate with a MWCO of 1000. These membranes should be thoroughly washed with deionized mineral water before use for removing any sort of preservatives or stabilizers.

8.3 EXPERIMENTAL PROCEDURE FOR OIL DEGUMMING PROCESS

The oil miscella needs to be prepared before passing through the different types of membranes employed for the degumming of oil (Desai et al., 2002). The oil miscella can be prepared by treating the crude oil sample with hexane @ 1:3 ratio, followed by stirring the sample for uniform dispersion, after this the sample is filtered through the filter paper and kept ready for the degumming process. The temperature needs to be maintained at 25°C to prevent the evaporation of the solvent. At first, permeation experiments need to be done, carried out in the UF test cell model no. 402. The specifications of the test cell are that it must include a stirred UF cell with a length of 15 cm and a diameter of around 7 cm with the ability of withstanding a pressure capability of 7 psi along with a cell capacity of 100 to 500 ml. The vessel bottom was placed on a flat sheet membrane with a diameter of 4.3 cm, which was further anchored by a porous disk, which stood on a circular base plate. By means of coupling, the plate was attached to the test cell. The air cylinder was connected to the top of the cell in order to collect the permeate; a conduit was placed on the bottom plate of the test cell. Above the membrane, turbulence was created by a magnetic stirrer bar in order to reduce concentration polarization. Experiments were run by charging 200 ml oil miscella into the cell. For degumming process of oil, the various optimization parameters used were pressure @ 50 psi, temperature of around 50°C and magnetic spin bar rotation @ 1000 rpm. The experiments were finished by assuring that half of the oil was recovered. To check the process efficiency, the phospholipid content of the feed, as well as the finished oil, were compared. The study carried out by Desai et al. (2002) demonstrated that the MWCO of the membrane formed in the feed played a predominant role in the separation of the phospholipids, which was noticed for all the types of membrane used in their studies. The reverse miscella formed in the feed can be taken into account while discussing process efficiency during operations. Based on their experiments, they projected that the use of ultrafiltration membranes like UF I & II exhibited a lower degree of phospholipid separation with less formation of reverse miscella in the feed system, while the other two membranes, *viz.* UF III and NF membranes, exhibited 95% phospholipids separation, projecting a higher process efficiency.

8.4 DEWAXING AND DECOLORIZATION OF OIL

Crude oil needs to be dewaxed in order to remove high molecular weight waxes. Otherwise, it will render the oil cloudy, which may reduce the aesthetic appeal in such a way that consumers won't buy the oil because of its appearance. This problem seems to be severe, particularly when we store the oil under cold conditions; during this time, the triglycerides, gums and waxes may separate from the oil as a solid residue; therefore, before marketing it is very much essential to dewax the oil. All edible

APPLICATIONS OF MEMBRANE TECHNOLOGY

oil, including groundnut oil, sunflower oil, rice bran oil and soybean oil, can be dewaxed before its consumption. This process is known as winterization. Generally, the process is carried out by passing the cooled crude oil through crystallizers where crystals will be formed and incorporated with the filtering aids, which helps in efficient filtration; then the oil is sent to the maturator where the crystals grow; the oil should be continuously cooled with the help of glycol water, which will be flown through the coils; then the oil is passed through the filter to remove waxes. They are then passed through hermetic pressure leaf filters for efficient removal; this process is quite conventional.

Nowadays, the dewaxing process is carried out by successfully employing membrane technology. It can also be carried out using a microfiltration process. Roy et al. (2014) developed a cost-effective ceramic-based microfiltration membrane, which was made out of clay-alumina. They carried out the trials at a pilot level using rice bran oil; they found that the rice bran wax as well as soap particles in the miscella aggregate with temperature changes. The researchers used a cross-flow membrane filtration process for the removal of wax; they observed that nearly 70–80% of acetone insoluble residue was recovered from the rice bran oil samples using a cross-flow membrane filtration process. The intensity of color was minimized to 50%, with oryzanol retention of up to 70%. The deacidification process of oil was carried out using a neutralizer like NaOH up to 10%, which toned down the free fatty acid content to 0.2%. The time needed for carrying out the experiments was around 10 hours with a 0.7 bar trans-membrane pressure, and permeate fluxes of 15 and 8 L/m^2 hours were obtained for the degumming, dewaxing and deacidification of rice bran oil. They concluded that using a ceramic-based microfiltration membrane leads to retention of micronutrient content, especially oryzanol @1.5%, with less oil loss @ 2.6%. They finally recommended that this method could be widely adopted to enhance the oil yield with more nutrients.

Manjula and Subramanian (2006) compared a membrane process along with a conventional process for degumming, dewaxing, deacidifying and decolorization of edible oils with and without using solvents. They found that the use of the membrane process had shown several advantages compared to the traditional method of refining oils. They observed that the use of an ultrafiltration membrane had led to the separation of phospholipids from the solvent, diluted as well as undiluted oil samples, with a high degree of oil flux. They explained that the use of such ultrafiltration could successfully dewax the hexane diluted oil, while the microfiltration membrane process can dewax undiluted oil efficiently without any further precooling steps during the refining process.

The deacidification process carried out for oil with the incorporation of alkali, followed by a membrane filtration process, removed the free fatty acid present in the oil. The use of a nonporous membrane had led to the production of oil with a light color, which suggests that the membrane removed most of the pigments, such as xanthophyll and chlorophyll. They finally concluded that the UF membrane was the best source of degumming and dewaxing operations, while the nonporous membrane could be used for the complete process, such as degumming, dewaxing and oil decolorization.

Manjula and Subramanian (2009) studied the efficiency of using nonporous membranes for the continuous degumming, decolorizing and dewaxing of crude rice bran oil. The study shows that the use of a nonporous membrane projected the rejection of phospholipids @ more than 99% in industrial and laboratory samples. The analysis of phosphorus content in rice bran oil processed using membrane techniques were around 2 ppm, indicating that hydratable and non-hydratable phospholipids were rejected by the membranes. Wax content was reduced to the extent of 40–50%; the color reduction was up to 55%. The results further advised that using a nonporous membrane may be an efficient way to refine the oil without any hindrance to its quality parameters and its acceptability level.

The efficiency of cross-flow microfiltration was studied by Pioch et al. (1998) in order to survey the effect of pore size, pressure and temperature on permeate flow rate as well as phosphatide and free fatty acid retention in rice bran oil. They added sodium hydroxide @ 5–40% to crude oil, and the filtration was carried out using a dead-end microfiltration with a pore size of 2.5 mm. The produced oil contained a phosphorus content of less than 5 ppm and a free fatty acid content of less than 0.1%, with a poor foul rate and membrane fouling; while the use of cross-flow filtration, i.e., using a tubular alumina membrane with a 12.0 mm pore size, showed a better permeability flow rate with reduced membrane fouling.

Suzuki et al. (1992) used an inorganic ultrafiltration membrane with a 30 mm diameter, a 0.2–10 µm thickness, and a pore size of 50 Å for removing phospholipids from vegetable oil. The operation temperature was 50–90°C, and the pressure was 2–5 kg/cm². They used hexane for extracting the oil from vegetable seeds, like sunflower, corn, soybean etc. These oils were passed through a microfiltration membrane made up of a sintered alloy, and the miscella solid materials were easily removed. Preheating of the oil hexane miscella was carried out and treated with ultrafiltration in two stages. They obtained a permeate phospholipid

APPLICATIONS OF MEMBRANE TECHNOLOGY

content of 20–30 ppm, a concentration of TAG @ 20–25% and the permeate flux was 130 l/m^2 h.

Hafidi et al. (2005) conducted a degumming cum deacidification process on vegetable oil using integrated membrane processing techniques. They couldn't detect monoglycerides, but the diglyceride content of oil was in the range of 0.8–1.0%. The sterol content of the oil was reduced to a greater extent.

The phospholipids were eliminated from the jatropha oil through an integrated approach of traditional degumming process cum UF membrane separation at a batch scale (Liu et al., 2012). They adopted a response surface methodology (RSM) to study the effects of temperature, speed of centrifugation process and the quantity of incorporated acid during the degumming process. They reported that the optimum processing temperature during the process was found to be 65°C with a centrifugal speed of 1600 rpm and acid addition @ of 4%. They showed that after a successful degumming process the phospholipid content was reduced from 1200 ppm to 60 ppm.

A degumming cum dewaxing process was carried out by filtration through a ceramic membrane followed by bleaching with earth material, and also refining was performed with the addition of alkali for crude rice bran oil (De and Bhattacharyya, 1998), and they compared this process with the traditional method, i.e., the centrifugal separation process. Their results showed that the use of a ceramic membrane enhanced the recovery of oil along with wax and gums. Finally, they concluded that the quality of the product was found to be highly satisfactory when using a ceramic membrane.

8.5 CONCLUSIONS

Ultrafiltration and microfiltration membranes can be widely used to dewax, degum and decolorize crude oil successfully. This process is environmentally friendly, and a higher yield can be obtained with less use of chemicals and without consuming much more energy. The oil obtained using this technique has found to possess good retention of nutrients with a higher yield and higher degree of consumer acceptability.

REFERENCES

De BK, Bhattacharyya DK (1998). Physical refining of rice bran oil in relation to degumming and dewaxing. *J Am Oil Chem Soc* 75(11): 1683–1686.

Desai NC, Mehta MH, Dave AM, Mehta JN (2002). Degumming of vegetable oil by membrane technology.*Indian J Chem Technol* 9: 529–534.

Hafidi A, Pioch D, Ajana H (2005).Membrane-based simultaneous degumming and deacidification of vegetable oils. *Innov Food Sci Emerg Technol* 6(2): 203–212.

Kale V, Katikaneni SPR, Cheryan M (1999). Deacidifying rice bran oil by solvent extraction and membrane technology. *J Am Oil Chem Soc* 76(6): 723–727.

Koseoglu SS, Lawhon JF, Lucas EW (1990). Membrane processing of crude vegetable oils: Pilot plant scale removal of solvent from oil miscella. *J Am Oil Chem Soc* 67(5): 315–322.

Liu K-T, Gao S, Chung T-W, Huang C-M, Lin Y-S (2012). Effect of process conditions on the removal of phospholipids from Jatropha curcas oil during the degumming process. *Chem Eng Res Des* 90(9): 1381–1386.

Manjula S, Subramanian R (2006). Membrane technology in degumming, dewaxing, deacidifying, and decolorizing edible oils. *Crit Rev Food Sci Nutr* 46(7): 569–592.

Manjula S, Subramanian R (2009). Simultaneous degumming, dewaxing and decolorizing crude rice bran oil using nonporous membranes. *Sep Purif Technol* 66(2): 223–228.

Pioch D, Larguize C, Graille J, Ajana H, Rouviere J (1998).Towards an effective membrane based vegetable oils refining. *Ind Crops Prod* 7(2–3): 83–87.

Roy B, Dey Surajit, Sahoo Ganesh C, Roy Somendra N, Bandyopadhyay Sibdas (2014). Degumming, dewaxing and deacidification of rice bran oil-hexane miscella using ceramic membrane: Pilot plant study 91(8): 1453–1460.

Subrahmanyam CV, Bhowmick DN (1999). Membrane degumming of rice bran oil. *J Oil Tech Assoc of India* 31: 193–196.

Subramanian R, Nakajima M, Yasui A, Nabetani H, Kimura T, Maekawa T (1999). Evaluation of surfactant-aided degumming of vegetable oils by membrane technology. *J Am Oil Chem Soc* 76(10): 1247–1253.

Suzuki S, Maebashi N, Yamano S, Nogaki H, Tamaki A, Noguchi A. (1992). Process for refining vegetable oil. Mitsubishi Kakoki Kaisha Ltd. Assignee US Patent 5,166,376.

9

Retention of Antioxidants by Using Novel Membrane Processing Technique

Rahul Shukla, Mayank Handa, and Aakriti Sethi

Contents

9.1 Introduction	212
9.2 Membrane Retention and Their Mechanisms	213
9.3 Salient Features of a Few Membranes Used for the Retention of Antioxidants	215
9.3.1 Polymeric Membranes	215
9.3.2 Inorganic Membranes	216
9.4 Retention of Anthocyanin	217
9.5 Retention of Poly Phenolic Compounds	219
9.6 Summary	222
Conflict of Interest	224
Acknowledgment	224
References	224

APPLICATIONS OF MEMBRANE TECHNOLOGY

9.1 INTRODUCTION

A substance that impedes the oxidation of the oxidizable substrate at a concentration lower than that of the oxidizable substrate is described as an antioxidant. The mechanistic approach behind the functioning of antioxidants is to prevent the deterioration of cell components which occur due to the involvement of free radicals in biochemical reactions. Lately, it has been observed that free radicals alter the functioning of biomolecules and, thus, the imbalance between the antioxidants and free radical species is principally involved in the pathology of a number of diseases. Antioxidants usually work by three main mechanisms:

- Antioxidant activity exhibited by certain enzymes acting against the free radicals
- Disruption of propagating chain reactions where free radicals react with the biomolecules and further produce other free radicals which target other molecules
- Proteins which tend to bind to the transition metals (Young and Woodside 2001).

Vitamin E and vitamin C have exhibited antioxidant activity by protecting against the free radical generated reaction mechanism. Principally, the antioxidant effect observed in vegetables and fruits arises because of the presence of flavonoids and phenolic acid, despite being a rich source of carotenoids, vitamin E and C. Also, consumption of a diet abundant in vegetables and fruits has shown to improve the oxidative stress in plasma. However, this improvement has not been correlated with the levels of carotenoids and tocopherol in the plasma (Bazinet and Doyen 2017; Conde et al. 2013; Hogan et al. 1998; Li and Chase 2010; Ulbricht 2013; Zeman and Zydney 2017; El-Abbassi et al. 2011).

Considering the number of diseases associated with oxidative stress, the oxidative damage caused in the atherosclerosis due to the lipid peroxidation of LDL provides a robust explanation for antioxidants being a safeguard in cardiovascular diseases. Higher intake of food which is rich in antioxidants reduces the risk of cardiovascular diseases. Also, it has been observed that all those patients suffering from Acute Respiratory Distress Syndrome (ARDS) had oxidative stress, which is known to be one of the major reasons behind the pathophysiology of the aforementioned disease, thus, exhibiting the significance of antioxidants in everyday life (Handa et al. 2019; Young and Woodside 2001). This chapter focuses on the membrane techniques which could be used in order to retain the levels of antioxidants in food items even after they have passed the stages of storage, packaging, distribution and processing.

9.2 MEMBRANE RETENTION AND THEIR MECHANISMS

Separation processes or retention techniques have been found to be valuable to fractionate, concentrate and recover a number of phenolic compounds from either products or by-products or from the waste produced through the processing of biomass using either aqueous media or alcoholic extract (Manach et al. 2004). The separation processes are driven hydrostatically by a pressure gradient, which allows the flow of solute molecules across the permselective membrane. On the basis of pore size and transmembrane pressure across the membrane, the separation processes could be classified as shown in Table 9.1 (Pandey and Rizvi 2009):

Lately, osmotic distillation and membrane filtration have been classified as one of the most interesting separation techniques as they allow the flow of solute molecules across the permselective membrane in high concentrations at atmospheric temperature and pressure. Both these processes involve the transfer of water vapors across the porous membrane, which is hydrophobic in nature from a region where the water vapor pressure is high to a region with low water vapor pressure. However, the difference is that in the latter technique the water vapor occurs due to the difference in temperature across the membrane, while in the former, it arises due to the difference in the activity of water molecules between the saline and feed solution (Liu and Wang 2016; Mohanty and Purkait 2011). All the above-mentioned separation techniques offer numerous

Table 9.1 Classification of Separation Processes on the Basis of Pore Size and Transmembrane Pressure (TMP)

S. No.	Membrane Process	Acronym	Pore size	Transmembrane pressure (bar)
1.	Microfiltration	MF	0.1–5 μm	1–10 bar
2.	Ultrafiltration	UF	0.5–100 nm	1–10 bar
3.	Nanofiltration	NF	0.5–10 nm	10–30 bar
4.	Reverse osmosis	RO	<0.5 nm	35–100 bar

advantages, such as a high efficiency in separation processes, uncomplicated equipment involved, ease in scale-up of the process and economical in terms of operating cost and maintenance (Ismail et al. 2015). Figure 9.1 and 9.2 represents the process of membrane distillation and osmotic distillation, respectively.

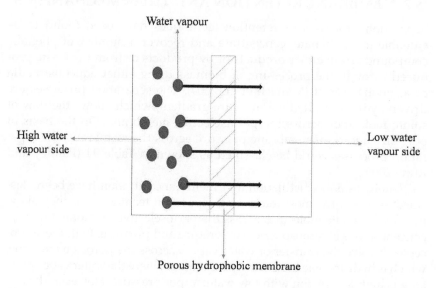

Figure 9.1 Diagrammatic representation of membrane distillation

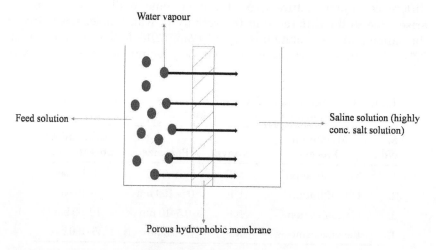

Figure 9.2 Pictorial representation of osmotic filtration.

9.3 SALIENT FEATURES OF A FEW MEMBRANES USED FOR THE RETENTION OF ANTIOXIDANTS

9.3.1 Polymeric Membranes

Despite the fact that a number of polymeric materials can be used for manufacturing the membranes involved in separation techniques, only a handful of them have been used for commercial purposes to be used as membranes for retention techniques (Vernhet and Moutounet 2002). The most commonly used polymers that have been employed on a commercial scale for membrane processes are discussed below.

1. Cellulose acetate (CA): Cellulose acetate polymers, belonging to the hydrophilic class of polymers, are composed of acetylated cellulose units. They are one of the most widely explored and extensively analyzed polymers, since first being used to prepare membranes which are anisotropic in nature. These membranes have exceptional flux capability for water molecules and rejection capability for salts (Van Der Bruggen et al. 2003).
2. Polysulphone (PS): Polysulphone polymers constitute aromatic and aliphatic units in sequence. Due to the presence of these sequences, polysulphone polymers have shown hydrophobic qualities and repel hydrophilic molecules. The surface of the membrane exhibits sparse hydrophilic characteristics by forming hydrogen bonds because of the aryl-SO_2-alkyl and aryl-O-alkyl linkage between the units. However, the presence of SO_2 molecules in polyethersulphone makes it comparatively less hydrophobic as the oxygen present in SO_2 binds with the water molecules.
3. Fluoropolymers (Poly(vinylidene difluoride)) (PVDF): These polymers are made up of aliphatic units, and the majority have a bond linkage between carbon and fluorine. They are highly hydrophobic because of their nonsusceptibility toward Van der Waals and hydrogen bonding.
4. Polytetrafluoroethylene (PTFE): These polymers are composed of monomers of tetrafluoroethylene, making them exceptionally hydrophobic in nature. They have been used in the preparation of porous membranes desirable for ultrafiltration and microfiltration through the application of heat and stretch.
5. Polypropylene (PP): These polymers are composed of monomers of propylene and show a high hydrophobic character. They are produced either using a thermal invasion technique or by melt-extrusion or stretching.

APPLICATIONS OF MEMBRANE TECHNOLOGY

6. Polyamide (PA): These polymers consists of an amide bond which links the repeating units. Examples of polymers with such a linkage include polyurethane, polybenzhydrazide, polybenzimidazole, nylon and polybenzamide. They are highly resistant toward oxidative agents and mechanical stress and, thus, highly used to prepare reverse osmosis or nanofiltration membranes.

7. Polyacrylonitrile (PAN): This polymer has been used for the production of RO and UF membranes and has also been shown to have extensive application in the textile industry. They have exhibited high resistance toward chemicals, i.e., a high stability toward chemicals, and have proven to be one of the highest-quality membranes employed in filtrations involving aqueous media.

In order to extend the performance characteristics of these membranes which are hydrophobic in nature, association of $ACOOH/ASO_3H$ with the surface molecules by chemical processes or photo-polymerization or addition of polymers which are hydrophilic in nature in the solution prepared for membrane manufacturing, for example, poly(vinyl methyl ether) and polyvinylpyrrolidone (PVP) are usually one of the alternatives used widely for industrial application (Van Gestel et al. 2002).

In contrast to all the above-mentioned polymeric membranes, thin-film composite membranes (TFC) have shown a relatively higher flux compared to the above-mentioned membranes. The latter consists of a dense supporting film that is microporous in nature, and which is covered by a thin layer of polymeric material. Hence, the produced membrane is anisotropic in nature with a thin overlying layer and a thick supporting basement layer. These membranes offer high resistance toward mechanical compression, high tolerability in pH values ranging from 2–11 and competent stability in terms of chemical and thermal stability (Cissé et al. 2011).

9.3.2 Inorganic Membranes

Ceramic membranes: Ceramic membranes, characterized as either an inorganic or a mineral membrane, exhibit various characteristic features such as:

1. High stability toward mechanical, chemical or thermal stress.
2. Inert toward the degradation caused by microbes.
3. Ease in cleaning in comparison to organic membranes.

Apart from the number of advantages offered, these membranes are quite expensive because of their required thickness for resisting pressure changes. The preparation of these membranes is carried out by combining carbides, oxides or nitrides of metals such as Ti, Zr and Al with supporting materials. Usually, ZnO_2, TiO_2, SiO_2 and Al_2O_3 are employed as base materials for preparing ceramic membranes (Quideau et al. 2011), generally in the range of NF or UF (Quideau et al. 2011).

Commercially, ceramic membranes are found in different forms, such as tubular (single/multichannel) or in the form of flat sheets. Possessing a composite design, these membranes are separated by a layer with a thickness of 1–2 mm, and a final layer, which is porous in nature (pore size = 10 mm), with a thickness of approximately 2 mm.

Usually, prepared ceramic membranes are operated easily at a transmembrane pressure of greater than 17 bars and at all pH values. Despite a number of advantages and its layer by layer design, these membranes have exhibited a restricted use compared to the wide variety of available polymeric membranes (Bazinet and Doyen 2017).

9.4 RETENTION OF ANTHOCYANIN

The occurrence of anthocyanin in juices, wines, flowers and fruits, makes them one of the most abundantly available polyphenols, which are hydrophilic in nature and are accountable for imparting pleasing shades ranging from the color red to violet. Apart from its ability to impart appealing shades, it has also been widely employed in the medicinal, nutrition and cosmetic industries, functioning as an antioxidant. Table 9.2 provides a brief description of the methods employed by various biological sources for the retention of anthocyanins. An ultrafiltration separation process was employed to extract lipase B in a purified form from *Candida antarctica* using an irregular and porous UF membrane. The use of such separation techniques for extraction circumvented the extensive and expensive stages employed in the purification of the enzyme.

As shown by Guimarães and group, the use of membrane technology led to a 2.5-fold improvement in the yield of obtained anthocyanins. In addition, the enzyme retained in the process made it possible to reuse it

Table 9.2 Various Filtration Process Employed for Retention of Anthocyanins From Various Biological Sources

S. No.	Source	Membrane Process	Membrane Configuration	Membrane Material
1.	*Candida antarctica*	UF	asymmetric	Supported by porous cellulose or PP/PE covered with PES
2.	Ethanolic jussara	NF	PES: smooth PA: rough, irregular	PES (NP030, MP010) PA(Desal 5-DK, Desal 5-DL, NF 270 and NF 90)
3.	Aronia solution	UF/NF	Flat sheets	Polysulfone
4.	Roselle extract	UF	Flat sheets	Polyethersulfone/ Composite PA

for at least three repeated reaction cycles with the same ester conversion degree (Manach et al. 2004).

Euterpe edulis, commonly known as jussara, has been widely employed for the ethanolic extraction of anthocyanins using a NF separation technique. The technique established by Vieira and group, utilized six membranes, which had molecular weights in the range of 150 to 1000 g/mol and were arranged in the form of flat sheets. The six sheets were assessed on the basis of the membrane texture and surface morphological characteristics and are classified below:

Out of all the six preferable membrane sheets (Figure 9.3), Desal 5-DK, purchased from GE Osmonics, exhibited the maximum capacity for

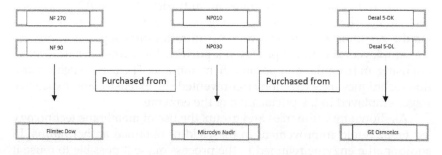

Figure 9.3 Diagrammatic presentation of preferable membrane sheets.

retaining anthocyanins because of high flux and small decline at 25°C and 25 bar pressure (Bravo 2009; Scalbert et al. 2005).

Fernando et al. carried out the separation of phenolic compounds in distinctive/varying proportions (W01WPM–W11WPM and W1FPC–W5FPC) using different levels of wine polymeric material. 3-AcGlc anthocyanins showed the highest capacity to retain itself, with their retention capacity varying from 0.77–0.87, which was found to be higher than the retention capacity of 3-Glc anthocyanins (0.47–0.52). Thus, the results suggested that the hydrophobic interactions between the anthocyanins and the polymeric material was responsible for a high value of retention capacity. The aforementioned study estimated that a gradual and continuous anthocyanin dose of red wine could provide a comparatively longer exposure to anthocyanins and, hence, provide further additional health benefits (Bendini et al. 2007).

Barber et al., demonstrated the retention of CAN obtained from Aronia solution by using a polysulphone membrane prepared using phase inversion. Determination of the influence of the polysulphone membrane on the retention ability of anthocyanins obtained from Aronia solution was carried out both in the presence as well absence of Na_2SO_4. Further, the study evaluated the retention capacity of Na^+ ions, other ionic compounds and anthocyanins. The presence of Na_2SO_4 almost led to a complete rejection of anthocyanins (more than 99%) during the separation process. Therefore, it could be concluded that the presence of Na_2SO_4 not only acts as a satisfactory preserve but also favorably inflates the effectiveness of the retention process (El-Abbassi et al. 2012; Visioli et al. 1999).

Hibiscus sabdariffa, commonly called roselle, has been evaluated for concentration of anthocyanins using a flat sheet membrane in the UF separation technique by Cisse and group. Anthocyanins were retained, ranging from the concentration of 24% (TMP of membrane = 5) with a membrane having a pore size of 150 KDa to 97% (TMP = 30) with a membrane of pore size 5 KDa. An increase in TMP value led to a high retention capacity of anthocyanins for all the membranes employed in the study. Thus, the results suggested that high levels of retention have been observed with the use of UF membranes because of the association between the material used for the membrane preparation and solute molecules (Conidi et al. 2015; Galanakis et al. 2010).

9.5 RETENTION OF POLY PHENOLIC COMPOUNDS

Originally found in vegetables and fruits, polyphenols are characterized as substances with a number of phenol groups (Paun et al. 2011). Polyphenols

APPLICATIONS OF MEMBRANE TECHNOLOGY

have been found to be beneficial because of their ability to impede the free radical mechanism and thus act as an antioxidant (Conidi et al. 2011). Based on the number of phenol groups present and the attached substituents which bind together a number of phenol rings, polyphenols are characterized in different classes (Borneman et al. 1997; Cassano et al. 2016). Table 9.3 provides a collection of various filtration techniques employed for the retention of polyphenolic compounds.

A number of membrane processes such as reverse osmosis, microfiltration and ultrafiltration have been employed for the isolation of the natural components present in olive mill wastewater. The presence of phenolic compounds of varying molecular weight have been found in olive mill wastewater, as shown in Table 9.4 (Sarmento et al. 2008; Todisco et al. 2002).

PS membranes have a 25 kDa molecular weight and a prominent efficiency for the isolation of pectin from the phenols. These membranes are also highly efficient in removing the hydroxycinnamic acid based derivatives as well as flavonols which have antioxidant properties (Al-Hassan et al. 2012; Servili et al. 2011).

Upon treatment of artichoke wastewater (ultrafiltered) using an UF membrane process with the help of spiral-wound polyamide membranes

Table 9.3   Various Methods of Filtration Employed for Retention of Polyphenolic Compounds

S. No.	Source	Membrane Process	Membrane Configuration	Membrane Material
1.	*Salvia officinalis* and *Geranium robertianum*	NF	Flat sheets	Polysulfone (PS) and PVP for making porous membrane
2.	Bergamot juice	UF/NF	Monotubular	PS
3.	*Morus alba*	UF/MF	Disk membrane	UK 200/50/10: PS UP 20: Polyamide (PA)
4.	*Castanea sativa*	UF	Flat membranes	Modified polyethersulphone
5.	Pequi	UF/NF	Flat sheets	UF: PS NF 90: PS and PA NF 270: PS and Polypiperazinea-mide

220

RETENTION OF ANTIOXIDANTS BY USING NOVEL MEMBRANE PROCESSING

Table 9.4 Phenolic Compounds Extracted from Olive Mill Wastewater

S. No.	Type of phenolic compounds extracted	Molecular weight of extracted phenolic compounds	Examples of the extracted compounds
1.	Low M. wt. phenolic compounds	Up to 198	Benzoic acid and its derivatives
2.	High M. wt. phenolic compounds	Up to 378	Secoiridoid aglycones
3.	Lignans	Up to 416	
4.	Glycosylated phenolic compounds	Up to 526 to 625	Oleuropein and verbascoside

(NF 270) (MWCO: 200–300 Da), a high concentration of phenolic (apigenin 7-Oglucoside and chlorogenic acid) as well as sugar compounds (glucose, sucrose and fructose) were extracted.

The NF membrane process has been widely employed for aqueous extraction of phenolic compound concentrate obtained from *Geranium robertianum* and *Salvia officinalis*. A high retention capacity for flavonoids and polyphenols has been observed with the use of membrane produced by dispersing PVP (20 g/L), PS (210 g/L) and SBA-15-NH2 (5 g/L) in N, N-dimethylformamide. A rejection of approximately 85.5 and 78.1% was observed for polyphenols concentrated from *Geranium robertianum* and *Salvia officinalis* respectively at a TMP of 10 bars (Guimarães et al. 2019).

Bergamot juice (ultrafiltered) has been studied comprehensively for retention of polyphenols using Titania membranes (MWCO: 450 Da and 750 Da) obtained from Inopor GmbH, Germany. With the use of an NF membrane with a pore size of 750 Da, a 44% rejection of polyphenols was observed. On the other hand, flavonoids obtained such as narirutin, neo-hesperidin, naringin and hesperidin, with a M.wt between 550 Da and 610 Da, exhibited rejection in the range of 43–62%. In contrast, with the use of a NF membrane with a pore size of 450 Da, rejection for flavonoids was quite high (~99%), and for sugar molecules it was observed to be approximately 48% (Benedetti et al. 2013).

Mulberry extract (*Morus alba*) obtained from the root or leaf, has been found to exhibit an antioxidant activity. However, due to its extensive extraction processes, the content of active compounds extracted is noted to be low. Thus, the use of membrane filtration techniques has shown comparatively better retention and a higher content of active compounds

APPLICATIONS OF MEMBRANE TECHNOLOGY

when compared to the conventional filtration techniques. The filtration process was performed at a temperature of 25°C with a TMP of 480.35 kPa using either 0.45 μm MF membranes or UF membranes with different cut off values for molecular weight such as UK 200/50/10 or UP 20. The concentrations of the active phenolic components, p-hydroxybenzoic acid and chlorogenic acid, obtained using this advanced membrane technique is comparatively higher than the conventional technique (Conde et al. 2013).

The use of 2 UF membranes (5 kDa and 10 kDa) in order to recover antioxidants from the leaves of *Castanea sativa*, observed an escalation in the phenolic concentration present, such as apigenin, rutin, 4-hydroxybenzoic acid, gallic acid, protocatechinic acid, vanillic acid and quercetin, by approximately 18%. Also, upon using ethanol in order to carry out further precipitation, an increase in the phenolic concentration by 36% was observed (Ulbricht 2013).

Extraction of glutathione and coumarins from pineapple juice, using either MF (pore size = 0.2 μm) or UF membranes, also improved the recovery of phenolic compounds to approximately 93%. Using ultra and nanofiltration membranes in order to extract lutein, xanthin and β-carotenoids from Pequi (*Caryocar brasiliense*), an improvement in the extracted phenolics concentration was observed when the filtration was performed at 25°C for UF (7 bars) and NF (8 bars) (Li and Chase 2010).

Further, below (Table 9.5) are the few listed examples of the separation (retention) and purification techniques of polyphenolic compounds using membrane technology.

9.6 SUMMARY

Indeed, hydrophobic and electrostatic interactions, adsorption phenomena, steric hindrance as well as solution effects on the membrane, solute/membrane properties and operating conditions play a key role for the removal efficiency and/or permeation of phenolic compounds through membranes. The membrane separation technique provides an edge over traditional conventional techniques because of the fact that it requires less energy, and the temperature used for separation is very low; thus, it is advantageous for thermolabile components. Hence, membrane separation techniques are being employed and will be widely used in the future as one of the best methods for the retention, extraction and purification of active components.

Table 9.5 Enlisted Examples of Purification and Retention of Polyphenolic Compounds Using Membrane Technology

Sources	Membrane processes	Membrane configurations	Membrane materials	References
Artichoke brines	NF	Spirally wounded	PPA, PA, PES	(Achour et al. 2012)
Apple juice	UF	Flat sheets	PVP, PES, RC	(Giacobbo et al. 2015; Vladisavljević et al. 2003)
Black tea leaves	UF	Tubular structure	Ceramic	(Mondal and De 2019; Shi et al. 2005; Sousa et al. 2016)
Cocoa seeds	RO, NF	Flat sheets	TFC, TFC/PA	(Albertini et al. 2015)
Olive mill wastewater	MF, UF, RO	Tubular, Spiral	PP, PA, TFC/PS	(Agalias et al. 2007; Takaç and Karakaya 2009)
Olive mill wastewater	MEUF	Flat sheets	PVDF	(Garcia-Castello et al. 2010; Mudimu et al. 2012)
Soybeans	NF	Spirally wounded	PVDF	(Hamzah and Leo 2018)
Thymus capitatus	NF, UF	Flat sheets	PA, CDA	(Jovanović et al. 2016; Jovanović et al. 2017)
Winery effluents	MF	Flat sheets, Hollow fiber	PVDF, Fluoropolymer, PI	(Czekaj et al. 2000; Ulbricht et al. 2009)

MF: microfiltration; **MEUF:** micellar enhanced ultrafiltration; **UF:** ultrafiltration; **NF:** nanofiltration; **RO:** reverse osmosis; **CDA:** cellulose diacetate; **PA:** polyamide; **PES:** polyethersulphone; **PVDF:** polyvinylidene fluoride; **PVP:** polyvinylpyrrolidone; **PS:** polysulphone.

APPLICATIONS OF MEMBRANE TECHNOLOGY

CONFLICT OF INTEREST

The authors declare no conflict of interest.

ACKNOWLEDGMENT

Authors acknowledge the Department of Pharmaceuticals, Ministry of Chemical and Fertilizers for financial assistance. NIPER-R communication number for this book chapter is NIPER-R/ Communication/091.

REFERENCES

Achour S, Khelifi E, Attia Y, Ferjani E, Noureddine Hellal A. Concentration of antioxidant polyphenols from thymus capitatus extracts by membrane process technology. *Journal of Food Science*. 2012.

Agalias A, Magiatis P, Skaltsounis AL, Mikros E, Tsarbopoulos A, Gikas E, et al. A new process for the management of olive oil mill waste water and recovery of natural antioxidants. *Journal of Agricultural and Food Chemistry*. 2007.

Al-Hassan AA, Norziah MH, Amorphopallus UI, Maizena DAN, Bertuzzi MA, Castro Vidaurre EF, et al. Improvement of bioactive phenol content in virgin olive oil with an olive-vegetation water concentrate produced by membrane treatment. *Food Chemistry*. 2012.

Albertini B, Schoubben A, Guarnaccia D, Pinelli F, Della Vecchia M, Ricci M, et al. Effect of fermentation and drying on Cocoa polyphenols. *Journal of Agricultural and Food Chemistry*. 2015.

Bazinet L, Doyen A. Antioxidants, mechanisms, and recovery by membrane processes. *Critical Reviews in Food Science and Nutrition*. 2017.

Bendini A, Cerretani L, Carrasco-Pancorbo A, Gómez-Caravaca AM, Segura-Carretero A, Fernández-Gutiérrez A, Lercker G. Phenolic molecules in virgin olive oils: A survey of their sensory properties, health effects, antioxidant activity and analytical methods. An overview of the last decade. *Molecules*. 2007.

Benedetti S, Prudêncio ES, Mandarino JMG, Rezzadori K, Petrus JCC. Concentration of soybean isoflavones by nanofiltration and the effects of thermal treatments on the concentrate. *Food Research International*. 2013.

Borneman Z, Gökmen V, Nijhuis HH. Selective removal of polyphenols and brown colour in apple juices using PES/PVP membranes in a single-ultrafiltration process. *Journal of Membrane Science*. 1997.

Bravo L. Polyphenols: Chemistry, dietary sources, metabolism, and nutritional significance. *Nutrition Reviews*. 2009.

Van Der Bruggen B, Vandecasteele C, Van Gestel T, Doyen W, Leysen R. A review of pressure-driven membrane processes in wastewater treatment and drinking water production. *Environmental Progress*. 2003.

Cassano A, Cabri W, Mombelli G, Peterlongo F, Giorno L. Recovery of bioactive compounds from artichoke brines by nanofiltration. *Food and Bioproducts Processing*. 2016.

Cissé M, Vaillant F, Pallet D, Dornier M. Selecting ultrafiltration and nanofiltration membranes to concentrate anthocyanins from roselle extract (Hibiscus sabdariffa L.). *Food Research International*. 2011.

Conde E, Reinoso BD, González-Muñoz MJ, Moure A, Domínguez H, Parajó JC. Recovery and concentration of antioxidants from industrial effluents and from processing streams of underutilized vegetal biomass. *Food and Public Health*. 2013.

Conidi C, Cassano A, Drioli E. A membrane-based study for the recovery of polyphenols from bergamot juice. *Journal of Membrane Science*. 2011.

Conidi C, Rodriguez-Lopez AD, Garcia-Castello EM, Cassano A. Purification of artichoke polyphenols by using membrane filtration and polymeric resins. *Separation and Purification Technology*. 2015.

Czekaj P, López F, Güell C. Membrane fouling during microfiltration of fermented beverages. *Journal of Membrane Science*. 2000.

El-Abbassi A, Khayet M, Hafidi A. Micellar enhanced ultrafiltration process for the treatment of olive mill wastewater. *Water Research*. 2011.

El-Abbassi A, Kiai H, Hafidi A. Phenolic profile and antioxidant activities of olive mill wastewater. *Food Chemistry*. 2012.

Galanakis CM, Tornberg E, Gekas V. Clarification of high-added value products from olive mill wastewater. *Journal of Food Engineering*. 2010.

Garcia-Castello E, Cassano A, Criscuoli A, Conidi C, Drioli E. Recovery and concentration of polyphenols from olive mill wastewaters by integrated membrane system. *Water Research*. 2010.

Van Gestel T, Vandecasteele C, Buekenhoudt A, Dotremont C, Luyten J, Leysen R, et al. Alumina and titania multilayer membranes for nanofiltration: Preparation, characterization and chemical stability. *Journal of Membrane Science*. 2002.

Giacobbo A, Do Prado JM, Meneguzzi A, Bernardes AM, De Pinho MN. Microfiltration for the recovery of polyphenols from winery effluents. *Separation and Purification Technology*. 2015.

Guimarães M, Pérez-Gregorio M, Mateus N, de Freitas V, Galinha CF, Crespo JG, et al. An efficient method for anthocyanins lipophilization based on enzyme retention in membrane systems. *Food Chemistry*. 2019.

Hamzah N, Leo CP. Fouling evaluation on membrane distillation used for reducing solvent in polyphenol rich propolis extract. *Chinese Journal of Chemical Engineering*. 2018.

Handa M, Sharma A, Verma RK, Shukla R. Polycaprolactone based nano-carrier for co-administration of moxifloxacin and rutin and its in-vitro evaluation for sepsis. *Journal of Drug Delivery Science and Technology*. 2019.

Hogan P, Canning R, Peterson P, Johnson R, Michaels A. A new option : Osmotic distillation. *Chemical Engineering Progress*. 1998.

Ismail AF, Padaki M, Hilal N, Matsuura T, Lau WJ. Thin film composite membrane - Recent development and future potential. *Desalination*. 2015.

APPLICATIONS OF MEMBRANE TECHNOLOGY

Jovanovic A, Djordjevic V, Zdunic G, Savikin K, Pljevljakusic D, Bugarski B. Ultrasound-assisted extraction of polyphenols from Thymus serpyllum and its antioxidant activity. *Hemijska Industrija Chemical Industry.* 2016.

Jovanović AA, Đorđević VB, Zdunić GM, Pljevljakušić DS, Šavikin KP, Gođevac DM, Bugarski BM. Optimization of the extraction process of polyphenols from Thymus serpyllum L. herb using maceration, heat- and ultrasound-assisted techniques. *Separation and Purification Technology.* 2017.

Li J, Chase HA. Applications of membrane techniques for purification of natural products. *Biotechnology Letters.* 2010.

Liu Y, Membranes Wang G. Technology and applications. *Nanostructured Polymer Membranes.* 2016.

Manach C, Scalbert A, Morand C, Rémésy C, Jiménez L. Polyphenols: Food sources and bioavailability. *The American Journal of Clinical Nutrition.* 2004.

Mohanty K, Purkait MK. Membrane technologies and applications. *Membrane Technologies and Applications.* 2011.

Mondal M, De S. Purification of polyphenols from green tea leaves and performance prediction using the blend hollow fiber ultrafiltration membrane. *Food and Bioprocess Technology.* 2019.

Mudimu OA, Peters M, Brauner F, Braun G. Overview of membrane processes for the recovery of polyphenols from olive mill wastewater olive mill wastewater. *American Journal of Environmental Sciences.* 2012.

Pandey KB, Rizvi SI. Plant polyphenols as dietary antioxidants in human health and disease. *Oxidative Medicine and Cellular Longevity.* 2009.

Paun G, Neagu E, Tache A, Radu GL, Parvulescu V. Application of the nanofiltration process for concentration of polyphenolic compounds from geranium robertianum and Salvia officinalis extracts. *Chemical and Biochemical Engineering Quarterly.* 2011.

Puasa SW, Ruzitah MS. Sharifah ASAK. An overview of micellar - enhanced ultrafiltration in wastewater treatment process. *International Conference on Environment and Industrial Innovation.* 2011.

Quideau S, Deffieux D, Douat-Casassus C, Pouységu L. Plant polyphenols: Chemical properties, biological activities, and synthesis. *Angewandte Chemie - International Edition.* 2011.

Sarmento LAV, Machado RAF, Petrus JCC, Tamanini TR, Bolzan A. Extraction of polyphenols from cocoa seeds and concentration through polymeric membranes. *The Journal of Supercritical Fluids.* 2008.

Scalbert A, Manach C, Morand C, Rémésy C, Jiménez L. Dietary polyphenols and the prevention of diseases. *Critical Reviews in Food Science and Nutrition.* 2005.

Servili M, Esposto S, Veneziani G, Urbani S, Taticchi A, Di Maio I, et al. Improvement of bioactive phenol content in virgin olive oil with an olive-vegetation water concentrate produced by membrane treatment. *Food Chemistry.* 2011.

Shi J, Nawaz H, Pohorly J, Mittal G, Kakuda Y, Jiang Y. Extraction of Polyphenolics from plant material for functional foods - Engineering and technology. *Food Reviews International.* 2005.

Sousa dos LS, Cabral BV, Madrona GS, Cardoso VL, Reis MHM. Purification of polyphenols from green tea leaves by ultrasound assisted ultrafiltration process. *Separation and Purification Technology*. 2016.

Takaç S, Karakaya A. Recovery of phenolic antioxidants from olive mill wastewater. *Recent Patents on Chemical Engineering*. 2009.

Todisco S, Tallarico P, Gupta BB. Mass transfer and polyphenols retention in the clarification of black tea with ceramic membranes. *Innovative Food Science and Emerging Technologies*. 2002.

Ulbricht M. Introduction to membrane science and technology. Von H. Strathmann. *Chemie-Ingenieur-Technik*. 2013.

Ulbricht M, Ansorge W, Danielzik I, König M, Schuster O. Fouling in microfiltration of wine: The influence of the membrane polymer on adsorption of polyphenols and polysaccharides. *Separation and Purification Technology*. 2009.

Vernhet A, Moutounet M. Fouling of organic microfiltration membranes by wine constituents: Mportance, relative impact of wine polysccharides and polyphenols and incidence of membrane properties. *Journal of Membrane Science*. 2002.

Visioli F, Romani A, Mulinacci N, Zarini S, Conte D, Vincieri FF, Galli C. Antioxidant and other biological activities of olive mill waste waters. *Journal of Agricultural and Food Chemistry*. 1999.

Vladisavljević GT, Vukosavljević P, Bukvić B. Permeate flux and fouling resistance in ultrafiltration of depectinized apple juice using ceramic membranes. *Journal of Food Engineering*. 2003.

Young IS, Woodside JV. Antioxidants in health and disease Antioxidants in health and disease. *Journal of Clinical Pathology*. 2001.

Zeman LJ, Zydney AL. Microfiltration and ultrafiltration: Principles and applications. *Microfiltration and Ultrafiltration: Principles and Applications*. 2017.

10

Application of Membrane Processing Techniques in Wastewater Treatment for Food Industry

Rahul Shukla, Farhan Mazahir, Divya Chaturvedi and Vidushi Agarwal

Contents

10.1 Introduction	230
10.2 Membrane Application in the Food Industry for the Treatment of Wastewater	232
10.3 Wastewater of the Food Industry: Volume and Quality	233
10.4 Wastewater Management Techniques	234
10.5 Membrane Process Techniques	236
10.5.1 Microfiltration	236
10.5.2 Ultrafiltration (UF)	239
10.5.3 Ultrafiltration Systems	239
10.5.4 Nanofiltration	240
10.5.5 Reverse Osmosis	241
10.6 Management Techniques Used in Some Industries	244

APPLICATIONS OF MEMBRANE TECHNOLOGY

10.7 Conclusion	247
Conflict of Interest	247
Acknowledgment	247
References	247

10.1 INTRODUCTION

The major or key resources in the agro- or food-based industry are water, energy and raw materials. From these key ingredients, water is the most important as it is involved in every step of processing. Water is one of the major sources for cleaning the equipment and washing the raw materials; it is also required in many processes for sanitizing the areas or equipment involved in the process. Wastewater produced through food production is vastly variable, and contingent upon the different types of food processing procedures (e.g., organic product, vegetables, oils, dairy, meat, and fish), as shown in Figure 10.1.

Advances in innovation have indicated numerous methods for wastewater management generated during the various processes of the food industry. By proper segregation of various cleaning processes, the isolated constituents and uncontaminated water is frequently recyclable in a synthetically unaffected form, and an effective re-utilization can be accomplished. Advantages are achieved when one or both of the output watercourses from the membrane system are reused or re-utilized, therefore, decreasing overall procedures and limiting waste management costs. There is a need to focus on the advancement and utilization of membranes in wastewater treatment, especially in the food industry. Generation of wastewater varies in terms of volume and effluent discharge in the food and agro-based industries depending upon the source of raw material used. Discharge contains macropollutants that vary in their biochemical oxygen demand (BOD), chemical oxygen demand (COD), total suspended solids (TSS), nutrients, oils, fats etc. (Galanakis 2012). Above all, it also contains certain chemical-based substances that are used in the processing and essential for enhancing the long term life span of the packaged product (Oreopoulou and Russ 2007). Nowadays, technology advancements are not limited to just treating the wastewater to the extent that it can reduce the wastewater content before finally discharging it into sewage treatment plants (Fillaudeau et al. 2006; Thomas and Thomas 2017). Technology innovations now focus on improving the quality of the wastewater to such an extent that it can be reused again in the food processing industry. Further, many useful components get washed away in food

APPLICATION OF MEMBRANE PROCESSING TECHNIQUES

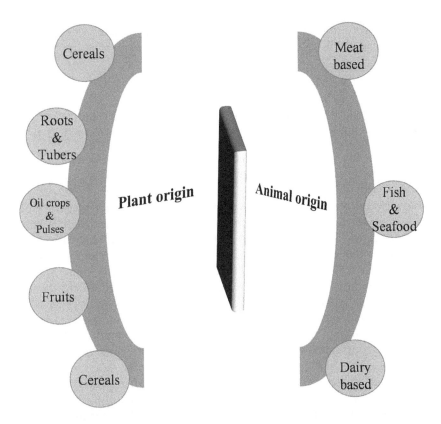

Figure 10.1 Various types of food waste and origin.

processing units, such as phenol-based compounds, certain proteins, vitamins etc., which can be collected and used for a more general purpose.

Conventional treatment is limited to three steps: primary treatment (i.e., physical and chemical treatment), secondary treatment (i.e., biological treatment) and tertiary treatment (i.e., activated carbon, precipitation etc.) (Ölmez and Kretzschmar 2009; Toczyłowska-Mamińska 2017). But with the advent of new technology, many innovative methods are not employed just for treating wastewater but also for recycling it and making it fit for reuse, and thus ending the possibility of a water crisis in the near future. In this chapter, a specific focus is given to the efficiency of membranes for wastewater treatment for recovery and different re-utilization purposes. It will provide an analysis of the engineering aspects along with the working conditions for improvements in wastewater management. A detailed

APPLICATIONS OF MEMBRANE TECHNOLOGY

point by point discussion is provided regarding elements of concern in water reuse applications, including its recovery in relation to the food industry.

10.2 MEMBRANE APPLICATION IN THE FOOD INDUSTRY FOR THE TREATMENT OF WASTEWATER

Membrane-based filtration procedures have a unique utility in the food industry as a wastewater treatment (Schlosser 2013). The procedure is utilized principally to lessen the bulk of the wastewater treatment, and this is accomplished with of two transitions: a saturate water motion with most of the first volume, and a concentrated motion with a lesser volume (constituents of effluents held) (Ortega-Rivas 2012; Qingyi et al. 2014). The membrane-based filtration utilized in wastewater treatment is broadly distributed in their structure and work. Basically, there are four types of membrane-based filtration: ultrafiltration (UF), microfiltration (MF), reverse osmosis (RO) and nanofiltration (NF) (Malik et al. 2013). Dissolvable porous material and selective partition are the two fundamental variables of these membranes. The different membranes installed in the transportation of wastewater decides the segregation of the molecular species at a particular stage. Molecule size is the sole measure and determines the saturation or dismissal of films. Microporous films (NF and RO) act as a barrier to the subatomic dimension and their selectivity is a barrier for the particular synthetic nature of the species (Belleville et al. 1992). Researchers across the globe have worked on these elements to clarify the detachment of deposits of residual wastewater. The effluent management techniques of the dairy industry by RO and NF films are shown in various researchers (Sarkar et al. 2006). However, a solid improvement in development and innovation can be seen with the outcomes in the other food industries. The food industry standard indicates that, procedure proposed for water reuse must equivalent to the rate of quality of drinking water. However, guidelines for other applications, for example, evaporator make-up water or warm cleaning water, are significantly stricter. There has been an investigation into the reprocessing of vapor condensate in milk handling operations (like dried milk powder production), such as kettle make-up water or the reuse of chiller spray water as warm cleaning water in meat handling operations.

232

For the most part, the pore stream or subatomic load of the molecule that is held or is separated by the layer make up the characterization parameters of the membranes. The membrane might have significant film properties, for example, a wettability surface, porosity, thickness, structure, and working conditions, and are additionally considered with the removal of solutes. The electrostatic repugnance between the film surface and waste product might be especially dissected to improve liquid-solute maintenance with an increment to water transition (Onsekizoglu 2012). The smallest molecular size present in the feed is significant for the determination of the layer's pore size estimate (Chapuis and Aubertin 2001). Feed characteristics, for example, pH modification, warm water treatment, expansion of synthetic concoctions, and prefiltration, could be altered before treatment. pH modification and warm treatment cause the precipitation of certain constituents.

10.3 WASTEWATER OF THE FOOD INDUSTRY: VOLUME AND QUALITY

Various food sources and their forms (e.g., natural products, vegetables, oils, dairy, meat, fish and so forth) change wastewater contaminants to a large extent. The qualities and rates of nourishment in wastewater are major factors and dependent upon the kinds of nourishment handling tasks, which include the wastewater management exercises in food cleaning (like purifying, stripping, cooking and cooling); precise exercises (like transport medium to transport food materials all through the procedure) and the cleaning of operational instruments between the various tasks (Bourgeois et al. 2001; Greenberg 1984; Strauss 1986). Likewise, one significant characteristic is the general size of the task, since food preparation extends from proceeding to next-level activities. Food preparation can be isolated into four noteworthy parts, i.e., meat, poultry and fish; foods grown from the ground and vegetables; refreshments and dairy. Table 10.1 summarizes the wastewater volume and contamination of some nutritious materials. Primary and secondary chemicals are regularly used to neutralize the high natural content of wastewater from the food industry using various methods and especially anaerobic technology (Gardner et al. 2012; Sahu and Chaudhari 2013).

After the regular treatment of wastewater, common prerequisites are secured by guidelines with more concentration on the cutoff points

APPLICATIONS OF MEMBRANE TECHNOLOGY

dependent on the shirking of contamination (Angelakis and Snyder 2015; Muga and Mihelcic 2008). Release licenses may incorporate stream, temperature, suspended solids, small molecular size solids, BOD, turbidity and amounts of nitrogen, and phosphorous. With an almost equivalent quality of treated wastewater to natural or stream water, it is transferred to the stream.

10.4 WASTEWATER MANAGEMENT TECHNIQUES

There are three main principle phases in the wastewater treatment cycle, commonly known as primary, secondary and tertiary treatment (Fu and Wang 2011; Qu et al. 2013; Trösch 2009). In certain applications, a further technique is required, known as a quaternary water treatment. This stage handles part per million to part per billion levels of contamination and often involves oxidation or fine filtration processes. Each of the mentioned stages deals with various poisons, while cleaning the water as it travels through the various stages (Kampschreur et al. 2009; Lakatos 2018; Rajkumar and Palanivelu 2004). Diverse treatment stages or groupings are adopted depending on the original quality of the water and its predetermined use (Osman 2014; Pokhrel and Viraraghavan 2004).

Primary wastewater treatment: During primary treatment, wastewater is incidentally stored in a settling tank where denser solids sink to the base, and lighter solids settle at the surface (Mara 2013; McCarty et al. 2011). When settled, these materials are kept down without any agitation, and supernatant fluid free of solid content is moved to the next step of the wastewater treatment. Large tanks are usually equipped with mechanical scrapers that continually keep the collected sludge in motion at the base of the tank and directs it to a hopper, which pushes it to the sludge treatment facilities.

Secondary wastewater treatment: Optional treatment of wastewater chips away at a more profound level than is essential, and it is intended to generously reduce the organic substance of the loss through natural oxygen consuming procedures. It is done in one of three different ways:

APPLICATION OF MEMBRANE PROCESSING TECHNIQUES

Table 10.1 Wastewater Generation with Various Average Parameters Generated from Different Food Processing Units

Feed processing	Unit/components	Wastewater generations	COD Amount	BOD Amount	References
Meat processing	Scalding tube	0.3	1800	1400	(Johns 1995; Sroka et al. 2004)
	Chiller showers	1.7	150	140	
	Cooling tanks	0.7	550	500	
Fruit juice	Orange	5.0	11,200	8100	(Caballero et al. 2015; Iaquinta et al. 2009)
	Apple	1.3	2000	1400	
	Tomato wastewater		1200		
	Fruit juice (general)		2500–7000		
Vegetable processing	Frozen carrots	30	5000	4500	(Charalabaki et al. 2005)
	Olive mill		100,000–200,000		(Olajire 2008)
Potato starch	Shower	0.7	3000	2500	
	Starch rinsing	1.5	7800	6500	
Beer production	–	4.2	2500	1800	
Alcohol plant	–	–	900–1200	–	
Fish industry	Unloading fish	–	5000–7000		(Cristóvaõ et al. 2015; Cristóvão et al. 2015; Matthiasson and Sivik 1980)
	Brine		4000–14,000		
	Cooked fish		4000–20,000		
Dairy industry	Whey	90	65000	42000	(Hung et al. 2005; Karadag et al. 2015)
	End pipe wastewater	1.5	1800	860	
	Flash cooling condensate		100570		
	Bottle rinsing		50–1000		
	Caustic solutions		8000–10,000		

APPLICATIONS OF MEMBRANE TECHNOLOGY

Bio-filtration: Bio-filtration uses sand channels, contact channels or streaming channels to guarantee that any extra residue is expelled from the wastewater (Srivastava and Majumder 2008; Tomar and Suthar 2011).

Air circulation: Air circulation is a protracted procedure that expands oxygen immersion by acquainting the air with the waste-water. Commonly, the air circulation procedure can keep going for as long as 30 hours, yet it is exceptionally powerful.

Oxidation lakes: Ordinarily utilized in hotter atmospheres, this strategy uses normal waterways, for example, tidal ponds, enabling wastewater to go through the process for a pre-decided period before being held for half a month.

Finishing optional wastewater treatment takes into consideration a more secure discharge into the neighborhood, diminishing normal biodegradable contaminants to safe levels.

Tertiary wastewater treatment: The essence of the tertiary treatment of wastewater is to raise the quality of the water to meet certain local and mechanical principles, or to fulfill explicit prerequisites around the sheltered release of water. Tertiary treatment likewise includes the expulsion of pathogens, which guarantees the potability of water. (Benford 1967; Lapara et al. 2011).

10.5 MEMBRANE PROCESS TECHNIQUES

Figure 10.2 provides an overview of a membrane enabled process used in filtration. There are four types of filtration, which are microfiltration (MF), ultrafiltration (UF), nanofiltration (NF) and reverse osmosis (RO).

10.5.1 Microfiltration

A physical partition is standard for the small scale filtration and ultra-filtration. The degree to which broken up solids, turbidity and microbes are expelled is controlled by the pore size of the membranes. Substances that are bigger than the pores in the membranes are completely expelled. Substances that are smaller than the pores of the membranes are halfway expelled, contingent upon the development of the decline layer on the film (Butcher 1990; Cheryan 1998). Small scale filtration and ultrafiltration are examples of pressure-dependent forms, which expel broken down solids

236

APPLICATION OF MEMBRANE PROCESSING TECHNIQUES

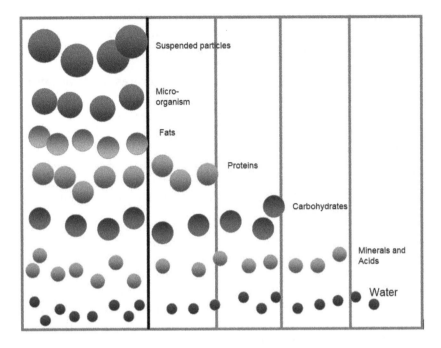

Figure 10.2 Diagrammatic representation of various size molecules passage through membrane pores.

and other substances from the water in comparison to nanofiltration and reverse osmosis.

Membrane Estimates in Microfilters

Membranes with pore sizes in the range of 0.1–10 µm are implemented in small scale filtration. Microfiltration membranes filter out all microorganisms. Microfiltration can be executed in a wide range of water treatment forms when particles with a width more prominent than 0.1 mm are to be expelled from a fluid (Abadi et al. 2011; Lim and Bai 2003).

Use of Microfiltration

Instances of small scale filtration applications are:

- Separation of microscopic organisms from the water (natural wastewater treatment).
- Effluent treatment.
- Separation of oil/water emulsions.

- Pretreatment of water for nanofiltration.
- Water processing and treatment of wastewater (Hatt et al. 2011; Wintgens et al. 2005).

Microfiltration is utilized widely in the water process and wastewater treatment:

- To reuse liquids utilized in tank/compartment cleaning.
- Clarifying the cleaning oils utilized in office printing and machine development.
- Removal of dangerous mixes in galvanic wastewater.
- Removal of suspended particles from material wastewater.
- Prefiltration for ultrafiltration, nanofiltration and turned around assimilation.

Procedures of Filtration

Microfiltration is a procedure of filtration with a micrometer measured channel. The channels can be in a submerged design or a weight vessel setup. They can be empty filaments, level, sheet, rounded, and winding. Figure 10.3 gives an overview of the microfiltration process. These channels are permeable and permit water as a monovalent species (Na+, Cl−), broken down natural tissue, little colloids and infections to go through. However, they don't permit particles, residue, green growth or enormous microscopic organisms through (Bandow et al. 1997; Gan et al. 1999).

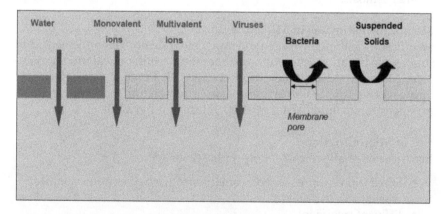

Figure 10.3 Schematic representation of the micro-filtration process.

10.5.2 Ultrafiltration (UF)

Utilization of Membrane Technology

- Membrane innovations give a significant and helpful solution for material wastewater treatment.
- Membranes help decrease contamination.
- As well as with reuse of important mixes from the waste streams.
- Help in the reuse of wastewater.
- Treatment of textile gushing using ultrafiltration.
- The material gushing treatment using a film procedure introduces some restrictions due to layer fouling, which causes fast transition decay.
- Pretreatment of the material by ultrafiltration.
- UF is viable as a single step treatment for optional material effluent (Banerjee et al. 2018; Chakrabarty et al. 2008; Shi et al. 2014).

Fundamental Constituents of Material Wastewater

The fundamental toxins in material wastewater are:

1. Organics
2. Hued colors
3. Toxicants
4. Inhibitory mixes
5. Surfactants
6. pH
7. Salts

10.5.3 Ultrafiltration Systems

Micro- and ultrafiltration systems can be offered for the treatment of efficient biological water treatment as an auxiliary system for evacuating infections and germs (Noble 2006; Zeman and Zydney 2017). They are likewise effectively utilized for biomass fixation. Extending the task time for the degreasing shower, phosphate shower and galvanic shower. Figure 10.4 provides an experimental setup of ultrafiltration.

Ultrafiltration and Its Uses:

Poly-ether-sulphone films are generally used for ultrafiltration (Madaeni et al. 2005). They have:

- Wide temperature limits

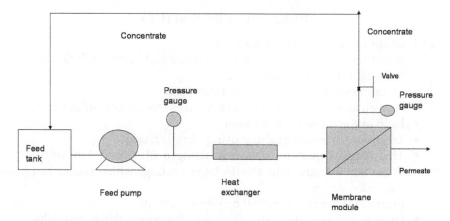

Figure 10.4 Systematic representation of experimental setup for ultra-filtration systems.

- Wide pH tolerance
- Good chlorine resistance
- Wide scope of pore sizes

Comparison with Customary Strategies

Ultrafiltration can be chosen instead of customary coagulation and sand filtration strategies for the decolorization of leftover color effluents (Baldasso et al. 2011).

- Ultrafiltration was found to be a suitable technique for the COD expulsion and decolorization of receptive colors and corrosive colors in the color profluent.
- Monetary evaluation demonstrated that in a sensible speculation the restitution time frame is less than six years

10.5.4 Nanofiltration

Progressively stringent water quality principles have prompted an expansion of productive film filtration advancements, similar to nanofiltration. Nanofiltration is, in some cases, used to reuse wastewater, as it offers higher transition rates and uses less vitality than an invert assimilation system (Van der Bruggen et al. 2008; Mänttäri et al. 2013; Mohammad et al. 2015). The plan and task of nanofiltration are fundamentally the same as that of turn around assimilation, with certain distinctions. The real distinction is that the nanolayer isn't as "tight" as the switch assimilation film (Pardhi et al. 2019; Shukla et al. 2019). It works at a lower feed water weight, and it doesn't evacuate monovalent (i.e., those with a solitary charge or valence of one) particles from the water as successfully as the RO layer (Oatley-Radcliffe et al. 2017; Petersen 1993). While a RO layer will regularly expel 98–99% of monovalent particles, for example, chlorides or sodium, a nanofiltration film commonly evacuates half to 90%, contingent upon the material and assembling of the layer. In view of its capacity to viably expel di- and trivalent particles, nanofiltration is often used to expel hardness from water while leaving more of the broken up solids content than would RO. Hence, it is known as the "conditioning film." Nanofiltration is regularly used to channel water with low measures of complete broken up solids, to evacuate natural substances and purify water (Bolong et al. 2009; Vandezande et al. 2008). Since it is a "looser layer," nanofiltration films are more averse to foul or scale and require less pretreatment than RO (Bolong et al. 2009). Here and there, it is even utilized as a pretreatment for RO. Nanofiltration can be utilized in an assortment of water and wastewater treatment businesses for the savvy evacuation of particles and natural substances (Van der Bruggen et al. 2003). Other than water treatment, nanofiltration is regularly utilized in the assembly procedure for pharmaceuticals, dairy items, materials and pastry kitchens. Figure 10.5 provides an overview of the nanofiltration process.

10.5.5 Reverse Osmosis

Reverse osmosis (RO) is an exceptional kind of filtration that uses a semipenetrable, dainty layer with pores small enough to let unadulterated water through while dismissing bigger particles, for example, disintegrated salts (particles), and different polluting influences, for example, microscopic organisms. Reverse osmosis is utilized to deliver exceedingly decontaminated water for drinking water frameworks, modern boilers, nourishment and refreshment preparation, beautifiers, pharmaceutical generation, seawater desalination and numerous different applications (Greenlee et al. 2009; Henthorne and Boysen 2015; Rao 2011). It has been a perceived innovation for over a century and popularized since the 1960s.

Advantages of RO:
- Easy to operate.
- Modular structure for simple establishment.

APPLICATIONS OF MEMBRANE TECHNOLOGY

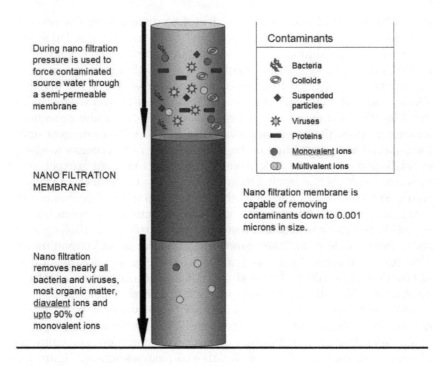

Figure 10.5 Workflow of the nanofiltration process.

- Does not require risky synthetic compounds.
- Reduces water and sewage use costs.
- Energy proficient, particularly when utilized rather than refining to deliver clear water.
- Can be coordinated with a current film filtration framework or particle trade framework to accomplish up to 80% flush water cycle.

About RO Membrane:
Reverse osmosis framework is working as an individual membrane. Every layer is a winding injury sheet of semi-porous material. Membranes are accessible in 2 inch, 4 inch and 8 inch width with the 4 and 8 inch sizes most utilized in industry (Cath et al. 2006; Lee et al. 2011; Malaeb and Ayoub 2011; Paul 2004). Industry has acknowledged a 40-inch length as a standard size so that films from various makers are compatible with gear frameworks. One of the essential estimations of a layer is its area. Membranes are accessible in the scope of 350–450 square feet of the surface

APPLICATION OF MEMBRANE PROCESSING TECHNIQUES

region. Semi-porous membranes were first developed utilizing cellulose acetic acid derivation (CAA) (Lee et al. 1981; Williams 2003). However, later industry changed to the utilization of a thin film composite (TFC) set over a more grounded substrate. TFC films are principally utilized today. Nowadays, RO treatment utilizes source water through a slim film and discrete pollutants from water (Arena et al. 2011; Ghosh et al. 2008). Figure 10.6 provides an overview of reverse osmosis.

RO works by switching the guidelines of assimilation, the normal inclination of water with broken down salts moves through the layer from a lower to higher salt fixation. This procedure is found all throughout nature. Plants use it to ingest water and supplements from the ground. In humans and other creatures, kidneys use assimilation to ingest water from the blood. In a RO framework, weight (more often than not from a siphon) is utilized to conquer the common osmotic weight, constraining the feed water with its heap of broken up salts and different polluting influences through a very complex semi-permeable film that evacuates a high level of the contaminations. The result of this procedure is exceptionally decontaminated water. The rejected salts and polluting influences

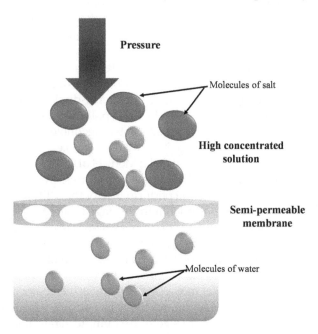

Figure 10.6 Basic principle of reverse osmosis.

APPLICATIONS OF MEMBRANE TECHNOLOGY

concentrate over the layer and are passed from the framework to deplete or onto different procedures. In an ordinary modern industry application, 75% of the feed water is filtered. In applications in which water protection is significant, 85% of the feed water is decontaminated. A RO framework uses cross-filtration, where the arrangement crosses the channel with two outlets: the separated water goes one way, and the tainted water goes another way (Duan et al. 2015; Senthil and Senthilmurugan 2016). To stay away from the development of contaminants, cross-stream filtration enables the water to clear away contaminant development and enough disturbances to keep the layer surface clean.

10.6 MANAGEMENT TECHNIQUES USED IN SOME INDUSTRIES

Wastewater from the Beverage-Based Industry and Its Management Techniques

The beverage- or, more commonly called, alcohol-based industry generates lots of wastewater, i.e., during the cleaning of fruit and its processing, tank cleaning along with its parts and the washing of bottles before packaging, which when taken together generates a lot of wastewater which is often discharged untreated into the sewage. This wastewater has high COD values and can labeled high strength wastewater. Due to its high nitrogenous and phosphorus content it must be treated before being discharged into a water body or municipal sewage, and to reduce the pollution levels. Researchers have developed new and advanced alternative technology in comparison to a conventional activated sludge-based treatment, i.e., membrane-based reactors (MBRs) (Heal 2014; Lin et al. 2012; Meng et al. 2017; Poddar and Sahu 2017). Before using MBR based cleaning procedures, pretreatment steps must be given so as to enhance the performance of the MBRs. Pretreatment processes must involve a conventional three-stage treatment.

Practically, MBRs are performed on a pilot scale in the food industry (Blöcher et al. 2002). Pure drinking water was obtained from this in a pilot plant study. Wastewater was collected from two or more medium enterprises and stored in an equalization tank and properly mixed; the removal of solid content from the water took place. Effluent is then passed via a MBR, which is equipped with a membrane filter, i.e., an immersed membrane filter, tubular based, with a pore size of 0.04 µm, to retain the particulate matter. Using this process, the total organic content (TOC) of the effluent

244

water reduces to 96% in the very first step of the treatment. A membrane filtration process is optimized by using a two-step nanofiltration process combined with ultraviolet disinfection. In the very first step, nanofiltration is conducted using spiral-based wound membrane modules with a specifically low rejection rate for sodium chloride, limiting it to only 55% and simultaneously reducing organic-based compounds to approximately 97%. The succeeding step, i.e., second and final step. is performed using membrane modules based on nanofiltration enabling a rejection rate of more than 90% for sodium chloride (Banvolgyi et al. 2006; Catarino and Mendes 2011; Labanda et al. 2009; Massot et al. 2008; Mondal and Wickramasinghe 2008). This type of filtration also causes the desalting of the wastewater collected and makes the water fit for drinking (Shon et al. 2013). When quality parameters were observed for COD and TOC, COD was not observed as it was below the detection limit and TOC was calculated to be 4 mg/L.

Wastewater from the Fish and Seafood-Based Industry and Its Management Techniques

The discharge of several effluents into wastewater during fish meal production, as well as rinsing, cleaning/thawing and cooking, gives the wastewater different characteristics. Pollution load and the discharge of effluents vary in accordance with the source from which it is collected and also the activity it undergoes. According to various literature reviews, salt content found in wastewater generated from this accounts for merely 0.25 to 10 g/L with COD in the range of 7 to 49, respectively (Afonso and Bórquez 2002; Osuna-Ramirez et al. 2017). One of the advantages of the wastewater generated from this industry is that it has no toxic effluents, but it still can't be discharged untreated in municipal sewage. Wastewater generated from the fish industry contains many valuable substances especially related to food segments, like various types of useful proteins, polyunsaturated fatty acids, flavors, aroma etc., which on recovery can yield highly productive gains. Membrane enabled technology is known to have potential benefits in relation to wastewater management, especially in fishing based industry. Afonso and Bórquez have successfully applied pressure based membrane filtration for treating wastewater generated from seafood-based processing units (Afonso and Bórquez 2003). They employed microfiltration, ultrafiltration or nanofiltration based membrane filtration techniques for recovery of the proteins extracted during the treatment of seafood-based industry processes. Pretreatment with microfiltration not only limited the oily and greasy content but also suspended particulate matter from the incoming wastewater. The ceramic membrane employed for ultrafiltration manages to reduce the organic content while enabling the recovery of valuable materials.

Dairy-Industry-Based Wastewater and Its Management

The dairy industry is one of the main producers of pollutants in the food industry in terms of quantity as well as quality. Wastewater generated from the dairy industry is examined using various parameters like pH, conductivity, COD, TSS, TOC and the residual form of hardness (Carvalho et al. 2013; Janczukowicz et al. 2008). Membrane-based recovery of wastewater along with essential compounds is a highly economical and beneficial concept with consideration for meeting global standards of nutritional demand as well as preventing a water shortage crisis. NF- and RO-based membrane filtration are very useful for concentrating milk-based constituents (Do et al. 2012; Sillanpää et al. 2014). Various researchers have documented the performance of membrane filtration techniques, which involved both NF and RO filtration for treating skimmed milk. The conventional method of drying whey for a smaller final product is now obsolete by big industries because it is expensive and it doesn't allow for extraction of other valuable components. For the extraction of valuable components from whey, membrane-based filtration was followed from past decades, yielding high inputs from industry with economic measures.

Wastewater from the Meat Industry and Its Management

The meat industry generates a large amount of wastewater with a huge amount of organic compounds, as well as salts, which are discharged directly into freshwater streams (Amin et al. 2014; Desmond 2006). The pretreatment of effluent discharge from industry involves flocculation, sedimentation, and coagulation as the primary treatment for the removal of solid substances as well as fat-based substances. Due to fat-based substances in the water, the growth of some microbial substances occurs, which involves the use of membrane-based filtration processes for removal and to re-utilize the water. Water filtration processes involve a combination of membrane filtration techniques, such as coagulation in combination with ultrafiltration, ultrafiltration followed by reverse osmosis or coagulation involving ultrafiltration followed by reverse osmosis. Combination techniques not only enhance the efficiency of the filtration rate but also save energy (Desmond 2006; Le-Clech et al. 2006; Sroka et al. 2004).

10.7 CONCLUSION

Recently, global concerns about water have given attention to water conservation and its re-utilization. Proper knowledge about the technicality and expertise of the water purification process must be clarified by industrial as well as public sector authorities. In this chapter, the authors summarized the various membrane filtration processes and their uses, finally leading to a discussion on limiting the burden on the environment caused by various food processing units. Membrane-based reactors have already been used in industry for successfully treating wastewater. Research is underway to increase efficiency by involving a combination or optimization of various parameters for reducing membrane fouling. In the future more attention must be diverted toward the involvement of membrane-based filtration in industry, so as to reduce the operational cost as well as enhance the performance and life expectancy of the membrane. Pressure derived membrane filtration processes integrated with other processes can prove to be useful approaches in the re-utilization of wastewater generated from food processing units in the near future. Workers across the globe have provided various case studies in small or medium enterprises in the food industries for gaining a key advantage over conventional methods in comparison to membrane-based methodologies. Now this the need of time and compulsion to promote growth in the membrane-based market in the future for decreasing the burden on water and proper re-utilization of wastewater.

CONFLICT OF INTEREST

The authors declare no conflict of interest.

ACKNOWLEDGMENT

Authors acknowledge the Department of Pharmaceuticals, Ministry of Chemical and Fertilizers for financial assistance. NIPER-R communication number for this book chapter is NIPER-R/ Communication/092.

REFERENCES

Abadi SRH, Sebzari MR, Hemati M, Rekabdar F, Mohammadi T. Ceramic membrane performance in microfiltration of oily wastewater. *Desalination*. 2011.

Afonso MD, Bórquez R. Review of the treatment of seafood processing wastewaters and recovery of proteins therein by membrane separation processes – Prospects of the ultrafiltration of wastewaters from the fish meal industry. *Desalination*. 2002.

Afonso MD, Bórquez R. Nanofiltration of wastewaters from the fish meal industry. *Desalination*. 2003.

Amin MT, Alazba AA, Manzoor U. A review of removal of pollutants from water/wastewater using different types of nanomaterials. *Advances in Materials Science and Engineering*. 2014.

Angelakis AN, Snyder SA. Wastewater treatment and reuse: Past, present, and future. *Water (Switzerland)*. 2015.

Arena JT, McCloskey B, Freeman BD, McCutcheon JR. Surface modification of thin film composite membrane support layers with polydopamine: Enabling use of reverse osmosis membranes in pressure retarded osmosis. *Journal of Membrane Science*. 2011.

Baldasso C, Barros TC, Tessaro IC. Concentration and purification of whey proteins by ultrafiltration. *Desalination*. 2011.

Bandow S, Rao AM, Williams KA, Thess A, Smalley RE, Eklund PC. Purification of single-wall carbon nanotubes by microfiltration. *Journal of Physical Chemistry B*. 1997.

Banerjee P, Das R, Das P, Mukhopadhyay A. Membrane technology. *Carbon Nanostructures*. 2018.

Banvolgyi S, Kiss I, Bekassy-Molnar E, Vatai G. Concentration of red wine by nanofiltration. *Desalination*. 2006.

Belleville MP, Brillouet JM, De la Fuente BT, Moutounet M. Fouling colloids during microporous alumina membrane filtration of wine. *Journal of Food Science*. 1992.

Benford WR. Water and wastewater engineering. *Journal of the Franklin Institute*. 1967.

Blöcher C, Noronha M, Fünfrocken L, Dorda J, Mavrov V, Janke HD, Chmiel H. Recycling of spent process water in the food industry by an integrated process of biological treatment and membrane separation. *Desalination*. 2002.

Bolong N, Ismail AF, Salim MR, Matsuura T. A review of the effects of emerging contaminants in wastewater and options for their removal. *Desalination*. 2009.

Bourgeois W, Burgess JE, Stuetz RM. On-line monitoring of wastewater quality: A review. *Journal of Chemical Technology and Biotechnology*. 2001.

Van der Bruggen B, Mänttäri M, Nyström M. Drawbacks of applying nanofiltration and how to avoid them: A review. *Separation and Purification Technology*. 2008.

Van der Bruggen B, Vandecasteele C, Van Gestel T, Doyen W, Leysen R. A review of pressure-driven membrane processes in wastewater treatment and drinking water production. *Environmental Progress*. 2003.

Butcher C. Microfiltration. *Chemical Engineer*. 1990.

Caballero B, Finglas P, Toldrá F. *Encyclopedia of Food and Health*. Academic Press: Cambridge, MA, 2015.

Carvalho F, Prazeres AR, Rivas J. Cheese whey wastewater: Characterization and treatment. *Science of the Total Environment*. 2013.

Catarino M, Mendes A. Dealcoholizing wine by membrane separation processes. *Innovative Food Science and Emerging Technologies*. 2011.

Cath TY, Childress AE, Elimelech M. Forward osmosis: Principles, applications, and recent developments. *Journal of Membrane Science*. 2006.

Chakrabarty B, Ghoshal AK, Purkait MK. Ultrafiltration of stable oil-in-water emulsion by polysulfone membrane. *Journal of Membrane Science*. 2008.

Chapuis RP, Aubertin M. A simplified method to estimate saturated and unsaturated seepage through dikes under steady-state conditions. *Canadian Geotechnical Journal*. 2001.

Charalabaki M, Psillakis E, Mantzavinos D, Kalogerakis N. Analysis of polycyclic aromatic hydrocarbons in wastewater treatment plant effluents using hollow fibre liquid-phase microextraction. *Chemosphere*. 2005.

Cheryan M. *Ultrafiltration and Microfiltration Handbook*. CRC Press: Boca Raton, FL, 1998.

Cristóvaõ RO, Botelho CM, Martins RJE, Loureiro JM, Boaventura RAR. Fish canning industry wastewater treatment for water reuse – A case study. *Journal of Cleaner Production*. 2015.

Cristóvão RO, Gonçalves C, Botelho CM, Martins RJE, Loureiro JM, Boaventura RAR. Fish canning wastewater treatment by activated sludge: Application of factorial design optimization. Biological treatment by activated sludge of fish canning wastewater. *Water Resources and Industry*. 2015.

Desmond E. Reducing salt: A challenge for the meat industry. *Meat Science*. 2006.

Do VT, Tang CY, Reinhard M, Leckie JO. Degradation of polyamide nanofiltration and reverse osmosis membranes by hypochlorite. *Environmental Science and Technology*. 2012.

Duan J, Pan Y, Pacheco F, Litwiller E, Lai Z, Pinnau I. High-performance polyamide thin-film-nanocomposite reverse osmosis membranes containing hydrophobic zeolitic imidazolate framework-8. *Journal of Membrane Science*. 2015.

Fillaudeau L, Blanpain-Avet P, Water Daufin G. Water, wastewater and waste management in brewing industries. *Journal of Cleaner Production*. 2006.

Fu F, Wang Q. Removal of heavy metal ions from wastewaters: A review. *Journal of Environmental Management*. 2011.

Galanakis CM. Recovery of high added-value components from food wastes: Conventional, emerging technologies and commercialized applications. *Trends in Food Science and Technology*. 2012.

Gan Q, Howell JA, Field RW, England R, Bird MR, McKechinie MT. Synergetic cleaning procedure for a ceramic membrane fouled by beer microfiltration. *Journal of Membrane Science*. 1999.

APPLICATIONS OF MEMBRANE TECHNOLOGY

Gardner M, Comber S, Scrimshaw MD, Cartmell E, Lester J, Ellor B. The significance of hazardous chemicals in wastewater treatment works effluents. *Science of the Total Environment*. 2012.

Ghosh AK, Jeong BH, Huang X, Hoek EMV. Impacts of reaction and curing conditions on polyamide composite reverse osmosis membrane properties. *Journal of Membrane Science*. 2008.

Greenberg AE. Advances in standard methods for the examination of water and wastewater. *Proceedings – AWWA Water Quality Technology Conference*. 1984.

Greenlee LF, Lawler DF, Freeman BD, Marrot B, Moulin P. Reverse osmosis desalination: Water sources, technology, and today's challenges. *Water Research*. 2009.

Hatt JW, Germain E, Judd SJ. Precoagulation-microfiltration for wastewater reuse. *Water Research*. 2011.

Heal KV. Constructed wetlands for wastewater management. *Water Resources in the Built Environment: Management Issues and Solutions*. 2014.

Henthorne L, Boysen B. State-of-the-art of reverse osmosis desalination pretreatment. *Desalination*. 2015.

Hung Y-T, Britz T, van Schalkwyk C. Treatment of dairy processing wastewaters. *Waste Treatment in the Food Processing Industry*. 2005.

Iaquinta M, Stoller M, Merli C. Optimization of a nanofiltration membrane process for tomato industry wastewater effluent treatment. *Desalination*. 2009.

Janczukowicz W, Zieliński M, Debowski M. Biodegradability evaluation of dairy effluents originated in selected sections of dairy production. *Bioresource Technology*. 2008.

Johns MR. Developments in wastewater treatment in the meat processing industry: A review. *Bioresource Technology*. 1995.

Jönsson AS, Trägårdh G. Ultrafiltration applications. *Desalination*. 1990.

Kampschreur MJ, Temmink H, Kleerebezem R, Jetten MSM, van Loosdrecht MCM. Nitrous oxide emission during wastewater treatment. *Water Research*. 2009.

Karadag D, Köroılu OE, Ozkaya B, Cakmakci M. A review on anaerobic biofilm reactors for the treatment of dairy industry wastewater. *Process Biochemistry*. 2015.

Kennedy MD, Kamanyi J, Salinas Rodríguez SG, Lee NH, Schippers JC, Amy G. Water treatment by microfiltration and ultrafiltration. *Advanced Membrane Technology and Applications*. 2008.

Labanda J, Vichi S, Llorens J, López-Tamames E. Membrane separation technology for the reduction of alcoholic degree of a white model wine. *LWT – Food Science and Technology*. 2009.

Lakatos G. Biological wastewater treatment. *Wastewater and Water Contamination: Sources, Assessment and Remediation*. 2018.

Lapara TM, Burch TR, McNamara PJ, Tan DT, Yan M, Eichmiller JJ. Tertiary-treated municipal wastewater is a significant point source of antibiotic resistance genes into Duluth-Superior Harbor. *Environmental Science and Technology*. 2011.

Le-Clech P, Chen V, Fane TAG. Fouling in membrane bioreactors used in wastewater treatment. *Journal of Membrane Science*. 2006.

APPLICATION OF MEMBRANE PROCESSING TECHNIQUES

Lee KL, Baker RW, Lonsdale HK. Membranes for power generation by pressure-retarded osmosis. *Journal of Membrane Science*. 1981.

Lee KP, Arnot TC, Mattia D. A review of reverse osmosis membrane materials for desalination-Development to date and future potential. *Journal of Membrane Science*. 2011.

Lim AL, Bai R. Membrane fouling and cleaning in microfiltration of activated sludge wastewater. *Journal of Membrane Science*. 2003.

Lin H, Gao W, Meng F, Liao BQ, Leung KT, Zhao L, et al. Membrane bioreactors for industrial wastewater treatment: A critical review. *Critical Reviews in Environmental Science and Technology*. 2012.

Madaeni SS, Rahimpour A, Barzin J. Preparation of polysulphone ultrafiltration membranes for milk concentration: Effect of additives on morphology and performance. *Iranian Polymer Journal (English Edition)*. 2005.

Malaeb L, Ayoub GM. Reverse osmosis technology for water treatment: State of the art review. *Desalination*. 2011.

Malik AA, Kour H, Bhat A, Kaul RK, Khan S, Khan SU. Commercial utilization of membranes in food industry. *International Journal of Food Nutrition and Safety*. 2013.

Mänttäri M, Van der Bruggen B, Nyström M. Nanofiltration. *Separation and Purification Technologies in Biorefineries*. 2013.

Mara D. Domestic wastewater treatment in developing countries. *Domestic Wastewater Treatment in Developing Countries*. 2013.

Massot A, Mietton-Peuchot M, Peuchot C, Milisic V. Nanofiltration and reverse osmosis in winemaking. *Desalination*. 2008.

Matthiasson E, Sivik B. Concentration polarization and fouling. *Desalination*. 1980.

McCarty PL, Bae J, Kim J. Domestic wastewater treatment as a net energy producer-can this be achieved? *Environmental Science and Technology*. 2011.

Meng F, Zhang S, Oh Y, Zhou Z, Shin HS, Chae SR. Fouling in membrane bioreactors: An updated review. *Water Research*. 2017.

Mohammad AW, Teow YH, Ang WL, Chung YT, Oatley-Radcliffe DL, Hilal N. Nanofiltration membranes review: Recent advances and future prospects. *Desalination*. 2015.

Mondal S, Wickramasinghe SR. Produced water treatment by nanofiltration and reverse osmosis membranes. *Journal of Membrane Science*. 2008.

Muga HE, Mihelcic JR. Sustainability of wastewater treatment technologies. *Journal of Environmental Management*. 2008.

Noble J. Ultrafiltration membranes. *Water and Wastewater International*. 2006.

Oatley-Radcliffe DL, Walters M, Ainscough TJ, Williams PM, Mohammad AW, Hilal N. Nanofiltration membranes and processes: A review of research trends over the past decade. *Journal of Water Process Engineering*. 2017.

Olajire AA. The brewing industry and environmental challenges. *Journal of Cleaner Production*.

Ölmez H, Kretzschmar U. Potential alternative disinfection methods for organic fresh-cut industry for minimizing water consumption and environmental impact. *LWT – Food Science and Technology*. 2009.

APPLICATIONS OF MEMBRANE TECHNOLOGY

Onsekizoglu P. Membrane distillation: Principle, advances, limitations and future prospects in food industry. *Distillation – Advances from Modeling to Applications*. 2012.

Oreopoulou V, Russ W. *Utilization of By-Products and Treatment of Waste in the Food Industry*. Springer: Boston, MA, 2007.

Ortega-Rivas E. Membrane separations. *Food Engineering Series*. 2012.

Osman M. Waste water treatment in chemical industries: The concept and current technologies. *Journal of Waste Water Treatment & Analysis*. 2014.

Osuna-Ramirez R, Alfredo Arreola Lizarraga J, Padilla-Arredondo G, Arturo Mendoza-Salgado R, Celina Mendez-Rodriguez L, Las Tinajas Carr, et al. Toxicity of wastewater FROM fishmeals production and their influence on coastal waters. *Fresenius Environmental Bulletin*. 2017.

Pardhi VP, Verma T, Flora SJS, Chandasana H, Shukla R. Nanocrystals: An overview of fabrication, characterization and therapeutic applications in drug delivery. *Current Pharmaceutical Design*. 2019.

Paul DR. Reformulation of the solution-diffusion theory of reverse osmosis. *Journal of Membrane Science*. 2004.

Petersen RJ. Composite reverse osmosis and nanofiltration membranes. *Journal of Membrane Science*. 1993.

Poddar PK, Sahu O. Quality and management of wastewater in sugar industry. *Applied Water Science*. 2017.

Pokhrel D, Viraraghavan T. Treatment of pulp and paper mill wastewater – A review. *Science of the Total Environment*. 2004.

Qingyi X, Nakamura N, Shiina T. Food industry. *Micro- and Nanobubbles – Fundamentals and Applications*. 2014.

Qu X, Alvarez PJJ, Li Q. Applications of nanotechnology in water and wastewater treatment. *Water Research*. 2013.

Rajkumar D, Palanivelu K. Electrochemical treatment of industrial wastewater. *Journal of Hazardous Materials*. 2004.

Rao SM. Reverse osmosis. *Resonance*. 2011.

Sahu O, Chaudhari P. Review on chemical treatment of industrial waste water. *Journal of Applied Sciences and Environmental Management*. 2013.

Sarkar B, Chakrabarti PP, Vijaykumar A, Kale V. Wastewater treatment in dairy industries – Possibility of reuse. *Desalination*. 2006.

Schlosser Š. Membrane filtration. *Engineering Aspects of Food Biotechnology*. 2013.

Senthil S, Senthilmurugan S. Reverse osmosis-pressure retarded osmosis hybrid system: Modelling, simulation and optimization. *Desalination*. 2016.

Shi X, Tal G, Hankins NP, Gitis V. Fouling and cleaning of ultrafiltration membranes: A review. *Journal of Water Process Engineering*. 2014.

Shon HK, Phuntsho S, Chaudhary DS, Vigneswaran S, Cho J. Nanofiltration for water and wastewater treatment – A mini review. *Drinking Water Engineering and Science*. 2013.

Shukla R, Handa M, Lokesh SB, Ruwali M, Kohli K, Kesharwani P. Conclusion and future prospective of polymeric nanoparticles for cancer therapy. *Polymeric Nanoparticles as a Promising Tool for Anti-Cancer Therapeutics*. 2019.

252

APPLICATION OF MEMBRANE PROCESSING TECHNIQUES

Sillanpää M, Metsämuuronen S, Membranes Mänttäri M. *Natural Organic Matter in Water: Characterization and Treatment Methods.* Butterworth-Heinemann: Oxford, UK, 2014.

Srivastava NK, Majumder CB. Novel biofiltration methods for the treatment of heavy metals from industrial wastewater. *Journal of Hazardous Materials.* 2008.

Sroka E, Kamiński W, Bohdziewicz J. Biological treatment of meat industry wastewater. *Desalination.* 2004.

Strauss SD. Wastewater management. *Power.* 1986.

Thomas O, Thomas MF. Industrial wastewater. *UV-Visible Spectrophotometry of Water and Wastewater.* 2017.

Toczyłowska-Mamińska R. Limits and perspectives of pulp and paper industry wastewater treatment – A review. *Renewable and Sustainable Energy Reviews.* 2017.

Tomar P, Suthar S. Urban wastewater treatment using vermi-biofiltration system. *Desalination.* 2011.

Trösch W. *Water Treatment. Technology Guide: Principles – Applications – Trends.* Springer: Berlin, Germany, 2009.

Vandezande P, Gevers LEM, Vankelecom IFJ. Solvent resistant nanofiltration: Separating on a molecular level. *Chemical Society Reviews.* 2008.

Williams ME. *A Review of Reverse Osmosis Theory.* EET Corporation and Williams Engineering Services Company, Inc. 2003.

Wintgens T, Melin T, Schäfer A, Khan S, Muston M, Bixio D, Thoeye C. The role of membrane processes in municipal wastewater reclamation and reuse. *Desalination.* 2005.

Zeman LJ, Zydney AL. Microfiltration and ultrafiltration: Principles and applications. *Microfiltration and Ultrafiltration: Principles and Applications.* 2017.

INDEX

A

Abbe Nollet, 47
Acute Respiratory Distress Syndrome (ARDS), 212
AEM, *see* Anion exchange membranes
AGMD, *see* Air gap membrane distillation
Air circulation, 236
Air gap membrane distillation (AGMD), 62
Anion exchange membranes (AEM), 60, 61, 120
Anthocyanin, retention of, 217–219
Antioxidants, 212, 215–217
ARDS, *see* Acute Respiratory Distress Syndrome
ARLA Food (Denmark), 83
Aromsa Besin Aroma ve Katkı Maddeleri Sanayi Ticaret A.Ş. (Kocaeli/Turkey), 83
Asymmetric membranes, 22, 68
Atlantic Seafood Ingredients Company (La Baule, France), 85

B

Bacillus licheniformis, 156
Bentonite, 133
Bio-filtration, 236
Bipolar membranes, 121, 122

C

CA, *see* Cellulose acetate
Cake filtration, 135
Candida antarctica, 217
Casein preparation, 162
Castanea sativa, 222

Cation exchange membranes (CEM), 60, 61, 120
Cellulose acetate (CA), 215
Cellulose nitrate, 3
CEM, *see* Cation exchange membranes
Ceramic membranes, 24, 217
CFMF, *see* Cross-flow microfiltration technique
Charged mosaic membranes, 22
Coca-Cola FEMSA (Veracruz, Mexico), 84
Colour and pigments, 141–142
Complete blocking, 135
Concentrate flow rate, 109
Concentration difference (Δc), 51, 52
Concentration-driven membrane processes, 30
    diffusion dialysis and dialysis, 31–32
    forward osmosis (FO), 31
    gas separation, 32–33
    pervaporation, 33
Concentration polarization, 11, 48, 188, 189
Cross-flow configuration, 27
Cross-flow filtration, 68, 101, 102
Cross-flow microfiltration (CFMF) technique, 173, 207
Crude oil, 202
    degumming of, 203
    dewaxing and decolorization of, 205–208

D

DCMD, *see* Direct contact membrane distillation
Deacidification process, 206, 207
    of fruit juices, 122–125
Deacidified juices, 125–126

INDEX

Dead-end filtration, 101
Decolorization, crude oil, 205–208
Degumming process
  crude vegetable oil, 203
  membranes used for, 203–204
Demineralization, 185
Depectinisation, 133
Depolymerising enzymes, 132
Dewaxing process, crude oil, 205–208
Diafiltered milk, 28
Diafiltration, 28
Dialysis, 31–32
Diffusion dialysis, 31–32
Direct contact membrane distillation
    (DCMD), 62
Direct osmosis, 31
Donnan effect, 104
Dynamic membrane fouling, 190

E

ED, *see* Electrodialysis
ED3C unit, 123, 124
EDBM2C, 122, 124
Electrically charged membranes,
    20, 22
Electric potential difference (ΔE), 51, 52
Electrodialysis (ED), 34–35, 60–61,
    110–111, 186, 187
Electrodialysis cell configurations,
    123, 124
Electrodialysis techniques,
    deacidification of fruit juices
  application of, 120
  bipolar membrane, 122
  deacidification process, 122–125
  effect on, physicochemical and
    organoleptic properties,
    125–126
  methods, preventing fouling of
    membranes, 126–127
  monopolar membranes, 121–122
  overview of, 119–120
Electroosmosis, 34–35
Electrostatic repugnance, 233

Electro-ultrafiltration system, 145
ESL, *see* Extended shelf life milk
Euterpe edulis, 218
Extended shelf life (ESL) milk, 173
    preparation of, 160–162

F

Feed flow rate, 109
Filtration membranes, 23
Flat plate configuration, 25
Flow configuration, 27
Fluoropolymers (poly(vinylidene
    difluoride)) (PVDF), 215
FO, *see* Forward osmosis
Food waste and origin, 231
Forward osmosis (FO), 31, 55–56

G

Gas separation, 32–33
Gelatin, 133
Geranium robertianum, 221
Gulgun Dairy Products (Nicosia/
    Cyprus), 83

H

Hazlewood Manor Brewery
    (Staffordshire, UK), 84
Heat-precipitated whey protein
    (HPWP), 177
Hibiscus sabdariffa, 219
HMOs, *see* Human milk
    oligosaccharides
Hollow fiber modules, 111, 112
Hollow fibre membranes, 25, 26
HPWP, *see* Heat-precipitated whey
    protein
Human milk oligosaccharides
    (HMOs), 174
Hybrid ultrasonic-membrane
    system, 146
Hydrophobic pervaporation, 175
Hyperfiltration, 29

INDEX

I

Inert purge pervaporation, 64
Inorganic membranes, 24, 67, 68, 70, 216–217
Intermediate blocking, 135
Ion-exchange membrane, 120

K

Kieselsol, 133

L

Laccase, 144
α-Lactalbumin, 171
β-Lactoglobulin, 171
Lactose, 172, 174, 175
    concentration and purification of, 184–185
Liquid membranes (LM), 22–23
Loeb-Sourirajan membranes, 22
Looser layer, 241
Loose reverse osmosis, 29

M

Mass transport mechanisms, 5
MBR, *see* Membrane bioreactor technology
MBRs, *see* Membrane-based reactors
MD, *see* Membrane distillation
Mechanisms of pressure (ΔP), 51, 52
Membrana, 46
Membrane-based reactors (MBRs), 244
Membrane bioreactor (MBR) technology, 4, 29, 65–66
Membrane configuration, 24–27
Membrane distillation (MD), 33, 35–36, 61–63
    diagrammatic representation of, 214
Membrane filtration, 213
Membrane fouling, 7, 8, 11, 48, 105, 134, 135, 189–191
Membrane geometry, 25

Membrane materials, 23–24
Membrane modification, 49
Membrane modules, 111–112
Membrane processing techniques;
        see also individual entries
    advantages, 8–11, 47–49
    application of, 71, 83–85
    classes of, 13–18
        asymmetric membranes, 22
        configuration, 24–27
        liquid membranes (LM), 22–23
        materials, 23–24
        symmetric membranes, 18–22
    commercial applications of, 12–13
    concentration-driven membrane processes, 30–33
    diffusion dialysis and dialysis, 31–32
    disadvantages, 47–49
    electrodialysis (ED), 34–35, 60–61
    electroosmosis, 34–35
    for food materials, 69–71
    forward osmosis (FO), 31, 55–56
    gas separation, 32–33
    hybrid membrane process, earlier studies on, 72–82
    limitations, 11–12
    mechanisms of, 49–51
    membrane bioreactor (MBR), 65–66
    membrane distillation, 35–36, 61–63
    microfiltration (MF), 27–29, 56–57
    nanofiltration (NF), 29–30, 59–60
    overview of, 2–4, 46–47
    pervaporation (PV), 33, 63–65
    pressure-driven membrane processes, 20, 21, 27–30
    reverse osmosis, 29–30, 53–55
    separation, operating principle of, 5–8
    structure, geometry, and selectivity of, 66–69
    ultrafiltration (UF), 27–29, 57–59
Membrane processing techniques, fruit juices and wine using

257

INDEX

classical clarification procedure, 133
effect
    on permeate flux, 136–139
    on quality of juices and wines,
        139–143
microfiltration (MF), 134
novel approaches, 145–147
overview of, 130
pre-treatment effect on, 143–145
problems of, 134–135
turbidity sources, 130–133
ultrafiltration (UF), 134
Membrane retention, 213–214
Membrane separation, 169, 173, 222
    operating principle of, 5–8
    uses, 70, 168
MF, *see* Microfiltration
Microfilters
    membrane estimates in, 237
    procedures of, 238
    uses of, 237–238
    the water process and wastewater
        treatment, 238
Microfiltration (MF), 27–29, 56–57,
        100–103, 236–238
    engineering aspects of, 157–158
    equipment cleaning, 158–159
    in fruit juices and wine, 134, 138
    operational procedure, 158
    overview of, 156–157
    uses of, 159
        casein preparation, 162
        cheese processing
            industries, 162
        extended shelf life (ESL) milk
            products preparation,
            160–162
        whey protein separation, 162
Microorganisms, 140
Microporous films, 232
Microporous membranes, 18–20
Molecular weight cutoffs (MWCOs),
        58, 178, 182, 184, 204, 205
Monopolar membranes, 20, 121–122
MWCOs, *see* Molecular weight cut-off

N

Nanofiltration (NF), 4, 29–30, 59–60,
        103–106, 240–241
Neosepta AXE01, 123
Neosepta BP-1, 123
Nernst-Plank equation, 105
Nestle Waters (Surat Thani,
        Thailand), 84
NF, *see* Nanofiltration
Non-porous dense membranes, 20
Normal-flow filtration, 68–69
Novel membrane processing
        technique
    membrane retention and
        mechanisms, 213–214
    membranes features, used for
        antioxidants retention
        inorganic membranes, 216–217
        polymeric membranes, 215–216
    overview of, 212–213
    retention of
        anthocyanin, 217–219
        poly phenolic compounds,
        219–222
Numerous separation
        technologies, 169

O

Oil degumming process, 204–205
Oil miscella, 204
Oil refining, disadvantages of
        conventional methods of, 203
Olive mill wastewater, 223
    phenolic compounds extracted
        from, 220, 221
Osmotic distillation, 213
Osmotic filtration, 214
Oxidation lakes, 236

P

Palm oil mill effluent (POME), 85
Pectic substances, 131

INDEX

Pectinolytic enzyme, 144
PEEKWC membranes, 138–139
Permeate, 168
    flow rate, 109
    flux profile, 136
Pervaporation (PV), 4, 33, 63–65
Phenolics, 142–143
Phospholipids, 208
Plate modules, 111, 112
Polyacrylonitrile (PAN), 138, 216
Polyamide (PA), 216
Polymeric membranes, 215–216
Polymeric microporous membranes, 36
Poly phenolic compounds
    purification of, 222, 223
    retention of, 219–223
Polyphenols, 132
Polypropylene (PP), 215
Polysulphone (PS), 138, 215
Polytetrafluoroethylene (PTFE), 215
Poly(vinylidene fluoride) (PVDF)
        membranes, 139, 172
Polyvinylpyrrolidone (PVP), 139
POME, *see* Palm oil mill effluent
Pore-flow model, 103, 104
Porifera Inc. (USA, California), 83
Pressure-driven membrane
        processes, 20, 21, 27
    microfiltration and
        ultrafiltration, 27–29
    nanofiltration and reverse
        osmosis, 29–30
Primary wastewater treatment, 234
Procyanidins, 132
PS, *see* Polysulphone
PTFE, *see* Polytetrafluoroethylene
PV, *see* Pervaporation
PVDF, *see* Poly(vinylidene fluoride)
        membranes
PVP, *see* Polyvinylpyrrolidone

R

Response surface methodology
        (RSM), 208

Retention
    of anthocyanin, 217–219
    of poly phenolic compounds,
        219–222
Reverse osmosis (RO), 29–30, 53–55,
        107–110, 241–244
RSM, *see* Response surface
        methodology

S

Salvia officinalis, 221
Saponifying enzymes, 132
Secondary wastewater treatment, 234
Soluble solids, 139–140
Spiral wound modules, 25, 26, 111, 112
Standard blocking, 135
Static membrane fouling, 190
Steric effect, 105
Symmetric membranes, 68
    electrically charged membranes,
        20, 22
    microporous membranes, 18–20
    non-porous dense membranes, 20
Synthetic organic polymers, 23

T

Tertiary wastewater treatment, 236
TFC, *see* Thin-film composite
        membranes
Theoretical approach
    electrodialysis, 110–111
    membrane modules, 111–112
    microfiltration, 100–103
    nanofiltration, 103–106
    overview of, 97–100
    reverse osmosis, 107–110
    ultrafiltration, 106–107
Thermally-driven membrane
        processes, 35–36
Thin-film composite membranes
        (TFC), 22, 68, 216, 243
TMP, *see* Transmembrane pressure
Total organic content (TOC), 244